SUSTAINABLE DEVELOPMENT IN CENTRAL ASIA

SUSTAINABLE DEVELOPMENT IN CENTRAL ASIA

Edited by
Shirin Akiner, Sander Tideman
and
Jon Hay

St. Martin's Press
New York

St. Martin's Press, Scholarly and Reference Division, 175 Fifth Avenue, New York, N.Y. 10010

First published in the United States of America in 1998 Printed in Great Britain

ISBN 0–312–21931–8

Library of Congress Cataloging-in-Publication Data

Sustainable development in Central Asia / edited by Shirin Akiner, Sander Tideman and Jon Hay.
p. cm.
Includes bibliographical references and index.
ISBN 0–312–21931–8 (cloth)
1. Sustainable development–Central Asia.
2. Sustainable development–Central Asia–Case studies. I. Akiner, Shirin. II. Tideman, Sander.
III. Hay, Jon.
HC420.3.Z9E57   1998
333.7'0958–dc21   98-34842
CIP

Contents

Foreword by Punsalmaagiin Ochirbat, former President of Mongolia   vii

Foreword by Kushok Bakula Rinpoche, Ambassador of India to Mongolia   ix

On the Need for a Unified Approach to Sustainable Development in Central Asia Sander G. Tideman, Council for Sustainable Development of Central Asia (CoDoCA)   xi

Contributors   xvii

Map   xix

List of Illustrations   xx

PART ONE: DEFINING THE REGION   1

1 Conceptual Geographies of Central Asia   3
  Shirin Akiner

PART TWO: THEORETICAL APPROACHES TO
SUSTAINABLE DEVELOPMENT   63

2 Development and Globalisation: Social, Psychological and
  Environmental Costs   65
  Helena Norberg-Hodge

3 The Shortcomings of the Classical Economic Model:
  Appropriate Economic Parameters Are Required for
  Sustainable Development of Central Asia   75
  Sander G. Tideman

PART THREE: CASE STUDIES   91

4   Sustainable Development: The Mongolian Experience   93
    Z. Batjargal

5   Amalgamating the Free Market Economy and
    Traditional Nomadic Society   104
    Alicia J. Campi

6   Lake Hovsgol – Selenge River Project   118
    Zane G. Smith

7   Environmental Sustainability, Development and Planning
    in Tibet   125
    Graham Clarke

8   Desertification in Western China   140
    Wang Tao

9   China-Australia Sheep Project, Xinjiang   150
    Frank B. Roseby

10 Environmental Problems in Kazakhstan   159
    Aliya S. Beisenova

11 Sustainable Mineral-Driven Development in Turkmenistan   167
    Richard M. Auty

12 The Demographic Boom and its Impact on the Mountain
    Regions of Tajikistan   186
    Khojamakhmad Umarov

13 Promoting Integrated Mountain Development in the
    Hindu Kush-Himalayas   195
    Mahesh Banskota

14 The Rural Non-Farm Sector in India: Issues of Relevance
    to Development in Central Asia   208
    Thomas Fisher

Index   235

Foreword

Punsalmaagiin Ochirbat

President of Mongolia 1993–1997

First of all I would like to express my appreciation for the publication of this book which discusses the sustainable development of Central Asia, a region destined to play an important role in global political and socio-economic life, and to wish the Council for the Sustainable Development of Central Asia every success in its activities.

The early 1990s witnessed nations starting to redefine the principles of their interaction to harmonise them with the changing times, whereby a trend towards unification on a regional and subregional basis for the purpose of resolving security and developmental issues became a reality. Hence, organising at this juncture a conference on Central Asia's development issues and bringing together scholars and experts in the field for a competent exchange of views and ideas was an event not only of regional but also of global significance.

We, the Central Asian nations living in the heart of the Eurasian land mass, are all linked by many common traits determined by geography, our historical and cultural backgrounds, security interests and traditional relations. The level of our economic development is generally the same as well. The emergence of newly independent states on the world's political map following the disintegration of the Soviet Union has given birth to a new region with a political and economic identity of its own. It would be appropriate if this reality found due reflection in the fabric of the world community, in particular in the structures of the UN and other international bodies. I voiced this idea of mine at the UN General Assembly in 1992.

Today all countries of the region are carrying out structural changes in their economies in order to make a transition to market economy relations. They are living through highly similar difficulties to reach one and the same ends. Hence, our communality is further enhanced, clearly indicating that we have a host of issues, resolution of which calls for our concerted effort. This region is rich in energy-generating and other types of natural resources, as well as in intellectual potential which has not yet been fully tapped. Furthermore, in geopolitical terms it is a bridge that links Europe and Asia, and as such will play an important role in international relations. In this sense our region has bright prospects for development.

The first Conference on Sustainable Development in Central Asia was held in Mongolia. It was a brain-storming session designed to chalk out the directions and mechanisms of regional cooperation. Collaboration in order to create the legal and material prerequisites for access to the oceans and the all-Asian communications network would serve the interests of each and every one of us. We face a need and have a great possibility to develop integral security, economic and cooperation systems, and to work jointly to protect the environment and turn Central Asia into a stable, nuclear-weapon-free region.

To my mind, it would be proper if the Council for Sustainable Development in Central Asia came up with a set of recommendations on how we, as a region, can contribute to the international endeavour to ensure the right balance between human activities and Nature, that is, to ensure sustainable development. At this period of structural adjustment, environmental protection is an issue which calls for cooperation, as it is critical to developmental prospects and the future of our world. We could jointly devise an action plan for implementing the *Agenda 21* adopted by the Rio de Janeiro ecological summit of 1992, and submit a joint proposal on representing Central Asia on the Committee for Sustainable Development which was set up in the wake of this summit meeting.

I hope the conference will yield palpable results. Please, accept my wishes for every success in your work.

Foreword

Kushok Bakula Rinpoche

Ambassador of India to Mongolia

It is a great privilege to be asked to write a Foreword to a progressive work of this nature. The Council for the Sustainable Development of Central Asia pioneered a work of great significance by organising the 'Conference on Sustainable Development in Central Asia' in Ulaanbaatar, Mongolia, in 1994. Thanks to the dedication and organisational skills of those who worked with the Council, the conference was well attended and successful. Many thoughtful individuals interested in the region, from all walks of life, from Central Asia and other parts of the globe, shared their views, and generated a pool of farsighted ideas.

No breakthrough is worthwhile unless it is followed up. Therefore, the Council for the Sustainable Development of Central Asia is publishing this collective wisdom in the present book, for the benefit of one and all.

The book focuses on Central Asia, which is rich not only in natural resources but also in traditional lifestyles. These lifestyles have been developed through generations, in tune with local conditions, requirements and imperatives. Before replacing them, before rushing to give developmental prescriptions and transplanting alien developmental models onto Central Asia, there ought to be extensive debate on all aspects of the subject.

Recently I had the privilege of sharing the State of the World Forum, organised by the Gorbachev Foundation in San Francisco, with distinguished personalities from all over the world. Before that I was also able to attend the Summit of Religious Leaders, organised by the World Wildlife Fund, MOA Foundation of Japan and Pilkington Foundation, to discuss 'Religion and Conservation'. At both these fora, as at other fora of any significance, the right model of development and sustainable development were prominent themes.

These interactions and my own association with and observation of the Central Asian region for over five decades, in different places and in varying capacities, have thoroughly convinced me of the significance of holding such an extensive debate. This book makes a laudable attempt in this direction.

According to Lord Buddha's teachings, most of our suffering and misery originates from selfish striving for personal gain and disregard for the welfare of others. In such a situation, a spirit of cooperation and accommodation is not possible, which in turn hampers human progress. This work seeks to break this vicious circle, with stimulating presentations and analyses on what needs to be done for the collective good of the entire region. I commend this book to all those with an interest in Central Asia.

On the Need for a Unified Approach to Sustainable Development in Central Asia

Sander Tideman

Council for Sustainable Development of Central Asia

This book arose out of the first Conference on the Sustainable Development of Central Asia (CoDoCA) held in Ulaan Baatar, the capital of Mongolia, in September 1994. The 110 participants from about 25 countries, including the new republics of Central Asia – consisting of government officials, academics, representatives of non-governmental organisations and the private sector – met for three days to discuss issues of common concern to the region. The participants expressed concern about the ecologically fragile environment of this resource-rich region. They agreed on the need to discuss together the entire array of economic development problems and, for this purpose, the need to enhance a distinct Central Asian regional identity. China, Russia and India, who share important parts of Central Asia, sent significant delegations to the conference.

It was probably the first time ever that a conference on this topic had been held in Central Asia, hence most delegates felt that the event was of historic significance, with potentially far-reaching consequences. Yet the conference was born from a simple observation.

Central Asia, as a distinct region, has been brought to the attention of the wider world by momentous events such as the disintegration of the Soviet Union, the establishment of the new Central Asian states, economic reforms in China affecting the remote provinces of Xinjiang, Tibet and Inner Mongolia and the re-opening of old Central Asian trade routes. Old barriers have been dismantled, giving the world access to a largely unknown land with vast human and material potential, while the peoples within the region are becoming aware of their own common identity and rich cultural heritage. While for decades the region was regarded as a mere remote borderland by distant central governments, Central Asia is now re-emerging as a significant force in the international political and economic arena.

In view of its geographical location and unique environmental conditions, Central Asia is faced by limited opportunities to pursue social and economic development. In spite of their lack of political unity, the Central Asian lands share a large number of characteristics affecting their development potential, including:-

- land-locked location
- low density of population
- low degree of industrial development and poor infrastructure
- scarcity of cultivable land
- extreme climate
- rich mineral resources
- shortage of development capital

Most central governments in Central Asia have regarded these typical features as major obstacles on the way to development, which in this context has usually been defined in terms of concepts prevalent in Western industrial nations, such as industrial output, income per capita and gross national product. This notion of development, which is based on values that can be easily quantified and expressed in monetary terms, has been behind most development policies in the past decades. It led to the establishment of large capital- and energy-intensive centralised systems, which, however, could not easily – if at all – be sustained by local communities at the grass roots level. The dramatic shrinking of the Aral Sea, whose water was used to grow cotton for export, serves as a sad reminder of the failure by central planners to take local conditions into account.

Recognising some of the shortcomings of the traditional socialist model, the new Central Asian authorities are now implementing development strategies inspired by capitalism. Most significantly, most countries are replacing inward-oriented industrialisation strategies by export-oriented policies based on the model of the West and the Newly Industrialised Economies (NIEs) of East Asia, which has proved to be more effective in generating material wealth throughout the world.

Yet while the new policies constitute an improvement from a conventional economic point of view, their underlying ideology has not changed. By concentrating, like the Soviet policies, on efforts to boost central government statistics and to invest in a large-scale centralised infrastructure mainly to the benefit of urban elites, like the Soviet policies, they fail to appreciate Central Asia's typical socio-economic and environmental limitations. Though they have perhaps brought crude improvements in industrial efficiency and some aspects of resource exploitation, the majority of the current development policies, not surprisingly, do not seem to be providing the benefits which Central Asians expected from them. In fact, there is sufficient evidence to show that in many cases the new capitalist policies are having serious unbalancing long-term effects on the fragile ecology and social life of the region. The most prominent problem is the rapid spread of deserts. If this trend is not reversed, as one of the conference participants stated, 'Central Asia could well become the World's second Sahel'. Needless to say, such an environmental disaster at the heart of Asia would affect the global ecology as a whole.

A growing number of people are questioning the validity of the development policies currently implemented in the region, and are attempting to develop economic policies and technologies which are more appropriate to local conditions and conducive to environmental and social harmony. Among them are three initiators of the conference, each having expertise on different parts of Central Asia: Dr Shirin Akiner, an expert on the Islamic peoples of the former Soviet Union and Director of the Central Asia Research Forum of the School of Oriental and African Studies, University of London; Dr Ts. Batbayar, a Mongolian scholar of the Asian Pacific region and director of the Mongolian Institute of International and Oriental Studies of the Academy of Sciences; and myself, a student of China and Inner Asia for over a decade, and based in Beijing from 1990 to 1995 as representative of an international bank.

When we met on various occasions during 1992 and 1993, it was clear that we shared the belief that the abundant natural wealth of the Central Asian landmass and the traditional ecological way of living of many of the region's indigenous inhabitants are certainly not liabilities. On the contrary, they offer a real prospect of sustainable and equitable improvements in standards of living. This does however, presuppose that local resources are efficiently utilised and conserved, that traditional ways of living are improved instead of discouraged, and that appropriate economic parameters are applied. Because of their many common features, we feel that the people of Central Asia are able to share their experience and know-how. More often than not, social and economic successes from one region could be fruitfully transplanted to other parts of Central Asia.

Thus, thinking along these lines, the plan was born to bring together experts from the region and abroad to discuss these issues and to attempt to formulate a more appropriate development strategy for countries in Central Asia. A conference organising committee was formed under the auspices of the Mongolian Academy of Sciences, the Central Asia Research Forum of the School of Oriental and African Studies, University of London, and the Shambhala Foundation of the Netherlands.

In view of the novelty of our initiative and the wide range of subjects which need to be addressed in the context of the development of the huge landmass of Central Asia, we decided initially to invite participants from as many Central Asian nations as possible and from greatly varying backgrounds. The preparations were not without problems. Due to the sheer remoteness of Mongolia and limited transportation facilities in Central Asia, some important invitees were unable to attend. The invited participant from Tajikistan, for example, was prevented from participating in the conference due to the instability in his country. In the end we were fortunate in having an enthusiastic and colourful group of forty foreign people in Mongolia, many of whom were first-time visitors to the country, happily mixing among themselves and their hosts.

We received strong support from various Mongolian government officials, most importantly from the then President, Punsalmaagiin Ochirbat, who recognised the opportunity for Mongolia to play a leading role in the development of Central Asia, and who provided an opening address to the conference and a foreword to this book. Dr Z. Batjargal, Mongolian Minister for Nature and the Environment, not only spent considerable time on the preparations of the conference but also contributed an important key-note paper to the conference. We also received encouragement from Mr Jan Swietering, resident representative of the United Nations Development Program in Mongolia, who also kindly provided technical support.

We are indebted to HE Kushok Bakula Rinpoche, a Buddhist monk and diplomat from Ladakh in the Indian Himalayas and currently Indian Ambassador to Mongolia, who from the outset encouraged us to go ahead with our plan and contributed a preface to this book. He felt particularly strongly about the preservation of the traditional culture of Central Asia, having himself been educated in pre-war Ladakh and Tibet and been a frequent visitor to Mongolia and the Mongolian parts of the former Soviet Union for several decades. Another source of inspiration was Helena Norberg-Hodge, director of the International Society for Ecology and Culture and founder of the Ladakh Project in the Indian Himalayas which developed appropriate technologies that optimise local resources in harmony with nature and traditional society – technologies which could find wide application in Central Asia. Her presentation, which is critical of the conventional development model, struck a chord with many Mongolians.

Last but not least, we are grateful to the Commission of the European Union and the Australian International Development Assistance Bureau for providing timely financial support. The publication of this book has also been made possible by a grant from the European Union. Jon and Cybele Hay, of the Central Asia Research Forum, gave invaluable assistance in translating the Russian-language papers. The late Anthony Polsky, a former Asia correspondent of the *New York Times*, also kindly provided advice. Sadly, Graham Clarke, who contributed an insightful paper on his pioneering research work in Tibet is no longer with us either. He passed away unexpectedly while this book was being prepared for publication.

Preliminary versions of almost all the papers in this book were presented at the conference. There are two main themes. The first group of contributions are mainly concerned with exploring the question of an appropriate philosophical framework for sustainable development in the context of Central Asia. The second group presents case studies from different regions and fields of sustainable development.

It is interesting to note that different interpretations of the term 'sustainable development' can be found throughout the book. There seems to be agreement that sustainable development should be defined in terms of the carrying capacity of the ecosystem, and that sustainability represents an economic state where demands placed upon the environment by people and industry can be met without reducing the capacity of the environment to provide for future generations and other species. Yet there is disagreement and uncertainty about how to arrive at the state of sustainability. For some, the conservation of nature is both the starting point and the goal, not only for nature's sake but also because the traditional culture and in fact much of the national identity of the Central Asians is intrinsically connected to the natural environment. Others propose to redefine economic priorities while still taking into account current economic realities such as the need for energy and foreign exchange.

In any event, one is necessarily confronted with the dilemma *to what degree* natural resources can be exploited – for the sake of short-term economic benefits – without seriously compromising the earth's ability to sustain life in the long run. As one of the participants put it,

'How much of our natural water resources can be diverted to hydro-electricity development? Do we build irrigation systems for agriculture or dams for power generation?' Perhaps it was symptomatic of this uncertainty that a Mongolian delegate to the conference called upon CoDoCA for guidance on how Mongolia should implement United Nations' *Agenda 21* on sustainable development, to which the country had committed itself at the UN Environment and Development Conference in Rio de Janeiro in 1992. In response to requests like this, the delegates agreed to establish the Council for Sustainable Development of Central Asia (CoDoCA), so that the debate started in Mongolia could be continued in the future (CoDoCA has meanwhile been registered as a foundation in the Netherlands). It will be CoDoCA's task to help Central Asia on the way towards sustainable development.

At this stage no-one can offer clear-cut solutions to the wide array of problems that lie on the way. The CoDoCA conference, book and council could be regarded as a well-meant beginning to a dialogue on this immense topic. We do not wish to impose any of our views on the Central Asians. Naturally, each country and nation has the right to determine its own economic and social future. Still, we need to ensure that government officials and others involved in the development of Central Asia are well equipped for the important choices that lie ahead. The Western conventional development model may on the surface seem attractive to newly 'capitalist' states, but its inability to consider values that cannot be expressed by monetary weightings does not seem to make it appropriate to the reality of Central Asians in which unquantifiable environmental and social resources are so important. Even in the West there is a growing tendency to question the sustainability of our economic model, as environmental degradation continues, unemployment increases and social structures fall apart.

Central Asian countries face the monumental task of creating a development model more suitable for their particular social and natural environment. In the process they will need to reconsider their economic priorities in favour of what brings most benefit in the long run: harmony amongst their peoples and with nature. May this book be a drop in the ocean of wisdom required for this endeavour.

Contributors

Shirin Akiner is Director of the Central Asia Research Forum, School of Oriental and African Studies, University of London. She has studied the history and culture of Central Asia for many years, especially of the five former Soviet states. She has travelled extensively in the region. She has published a number of works on social and political change in these countries.

Helena Norberg-Hodge is founder and director of the Ladakh Project in the Indian Himalayas and the International Society for Ecology and Culture. She is the author of *Ancient Futures*, which describes her experience as witness of the rapid changes in the traditional life of Ladakh due to its sudden exposure to modern economics.

Sander G. Tideman trained in international economic law and finance, and worked for nine years in Taiwan, Hong Kong and China in international banking. He now works for the Triodos Bank, a Dutch-based bank committed to socially and environmentally responsible business. Among other projects, the Bank runs a micro-finance programme in Central Asia.

Dr Z. Batjargal was Mongolia's Minister for Nature and the Environment until 1996. He is now Director of the Mongolian Agency for Hydrometeorology and Environmental Monitoring and the Mongolia Environmental Trust Fund.

Alicia J. Campi is a business consultant devoted to Mongolia, China, Siberia and Central Asia, and a former president of the US-Mongolia Advisory Group. Dr Campi received a doctorate degree in Uralic and Altaic studies from Indiana University.

Zane G. Smith, Jr is Public Policy Coordinator of Ecologically Sustainable Development, Inc., Elizabethtown, New York, 12932, USA.

Dr Graham E. Clarke was, until his untimed death in 1998, Co-ordinator of Development and Anthropology and senior member of Queen Elizabeth House, University of Oxford.

Dr Wang Tao is Deputy Director of the Lanzhou Institute of Desert Research, Chinese Academy of Sciences, Lanzhou, China.

Frank B. Roseby was Australian Team Leader on the China-Australia Sheep Research project, Urumqi, Xinjiang Uygur Autonomous Region, PRC, from 1992 until 1996.

Dr Aliya S. Beisenova founded the Laboratory for the Conservation of Nature and the Ecology of Landscapes in the Department of Physical Geography at the Abay Pedagogical Institute, Almaty, Kazakhstan. This was one of the first ecologically focused university departments in the former Soviet Union.

Richard M. Auty is Reader in the Department of Geography at the University of Lancaster.

Dr Khojamakhmad Umarov was until recently a member of the Academy of Sciences of Tajikistan, Dushanbe; he now works as a freelance writer and consultant.

Mahesh Banskota is a development economist. He is currently Director of Programmes at the International Centre for Integrated Mountain Development (ICIMOD), Kathmandu, Nepal.

Thomas Fisher worked in India for five years on issues of rural livelihoods and of institutional development, including for economic development within the Tibetan community. In 1995 he documented and published on his Indian work as a Visiting Fellow at the Institute of Development Studies at Sussex University. He is currently at the New Economics Foundation, managing programmes on value-based organisations and on finance against poverty in India and Britain.

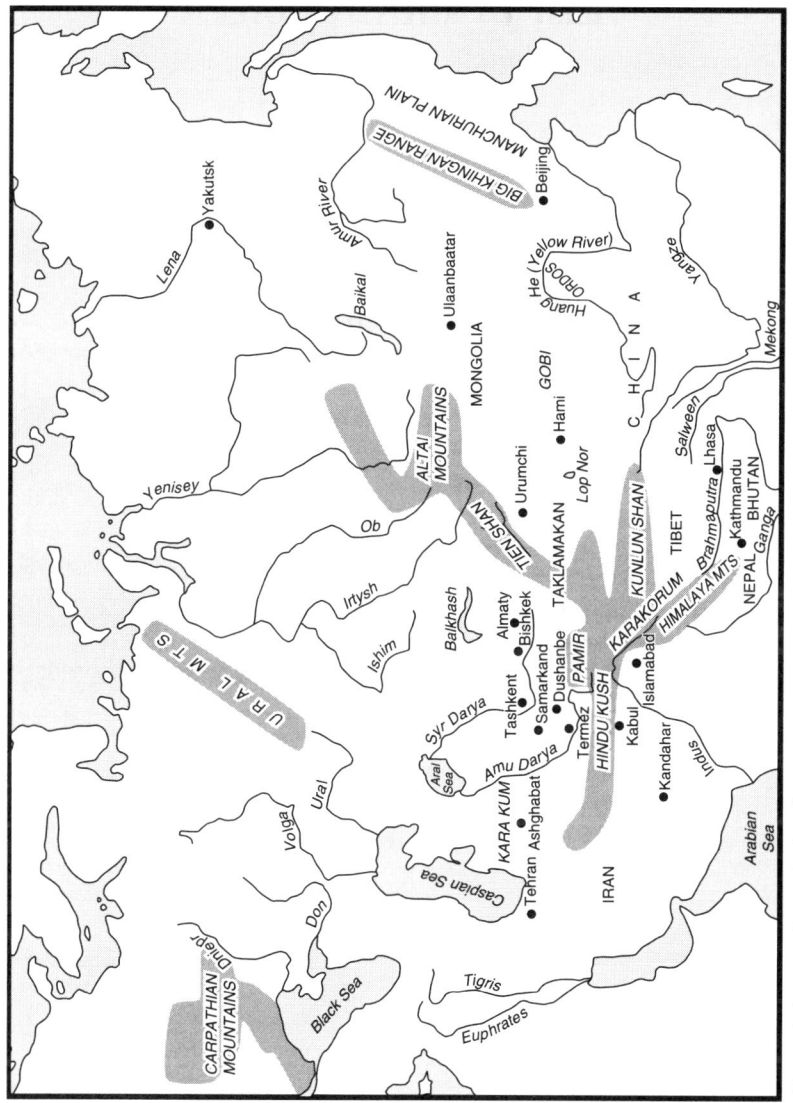

Central Asia and Borderlands

xix

List of Illustrations

Between pages 124 and 125.

1 Tibetan nomads, Sichuan province
2 Tibetan nomads, Sichuan province
3 Tibetan settlers, Sichuan province
4 The Muslim minority in Qinghai
5 Gardeners at work, Qinghai
6 Local fishing, Angara river, southern Siberia
7 Outskirts of Irkutsk with ring of dachas, southern Siberia
8 View of Lake Baikal
9 The Yellow River near Baotou, Inner Mongolia
10 Massive deforestation, Sichuan province
11 Trucks with Tibetan timber heading for Chinese factories
12 Water well, Mongolia
13 Nomads in northern Mongolia
14 Present day Ulaanbaatar, capital of Mongolia

PART ONE

DEFINING THE REGION

Chapter 1

Conceptual Geographies of Central Asia

Shirin Akiner

The title of this book specifies a regional focus, namely 'Central Asia'. At first glance this might seem to be a clear, unambiguous statement. However, as soon as one tries to set precise boundaries to the intended area of discussion the picture begins to blur. Exactly where and what is 'Central Asia'? And how does this term differ from the other terms that are used to refer to the same, or overlapping, or adjacent, parts of the Eurasian landmass? The aim of this chapter is, firstly, to try to shed some light on these problems of definition; and secondly, to discuss some of the historical linkages that are today contributing to a renascent sense of regional identity.

It is the shared perception of a body of common values, common experiences and common trajectories of development that give substance to the concept of a distinctive Central Asian 'footprint'. Awareness of a joint heritage has in recent years stimulated initiatives to create fora for intra-regional dialogue, consultation and cooperation. The present volume is the outcome of one such endeavour. The process has been given added impetus by huge changes in the political environment, most notably the disintegration of the Soviet Union and consequent emergence on its former southern border of five independent countries. This transformation of the political map has helped to bring new prominence to the region, but it has also made it important to clarify conceptual definitions of 'Central Asia', if only for the sake of greater mutual comprehension.

Given the profusion of terms that have been used to denote the region, and the various interpretations of these terms, this is no simple task. In an attempt to unravel the main sets of criteria that have been used, or could be used, as bases for marking out a coherent regional entity, this chapter reviews the key conceptual issues. Each has its own chronology, as well as 'geography' or spatial dimensions. In order to

3

highlight this, separate sections are devoted to each topic. This approach inevitably leads to a certain amount of repetition, but is perhaps less confusing than a composite treatment which tries to cover everything simultaneously. The final sections look at some of the common environmental hazards that afflict the region; they, too, constitute a conceptual geography of Central Asia – a 'geography of fragile environments' – and it is these fragile environments that form the subject of the remainder of the book.

Boundaries and Terminology

Modern terminology

There is no generally accepted definition of the term 'Central Asia'.[1] In English, as in most other western European languages, it has been variously used to describe, in whole or in part, a vast swathe of the Eurasian landmass. At its widest limits, the term may encompass a belt that stretches from the Hungarian plain in the west to the Ussuri and Amur rivers in the east, from the Arctic Circle in the north to the Indo-Gangetic plain in the south. More commonly, however, it is used to designate some part of the smaller (though nevertheless huge) area that is bounded by the Volga and Argun rivers, southern Siberia and the Himalayas.

In Russian (as also in languages that, for political reasons, were directly influenced by Russian usage), by contrast, especially during the Soviet period, there was a more precise definition. The term 'Middle Asia' (*Srednyaya Aziya*) was used to refer to the four southern republics of Kyrgyzstan, Tajikistan, Turkmenistan and Uzbekistan. Kazakhstan was regarded as a separate entity and hence referred to separately (thus, 'Middle Asia and Kazakhstan'). The term 'Central Asia' (*Tsentral'naya Aziya*) was reserved for the lands further to the east that were, or had been, under Chinese rule, namely, Mongolia (both the part formerly known as 'Outer' Mongolia, and Inner Mongolia), Xinjiang and Tibet (i.e. present-day Xizang/Tibet Autonomous Region, and adjacent areas of Chinese provinces with a significant Tibetan population). In the indigenous languages, terms to designate the region generally took local features as their points of reference (e.g. distinctive geographic markers, the dynastic name of the ruling state-tribe), or else adopted the terms used by the dominant imperial power.

Since the collapse of the Soviet Union, the terminology used in both Russia and the other former Soviet republics has undergone a change.

The five newly independent southern states (now including Kazakhstan) have adopted the term 'the Central Asia states' as their collective designation.[2] This has come to be the accepted international usage. In Russia likewise, *Tsentral'naya Aziya* is increasingly used in this way (at least with reference to the post-Soviet period), thereby removing the old lexical distinction between the lands within the Russian sphere of influence and those within the Chinese sphere of influence.

Historical usage

A number of terms have been used to designate parts of the Eurasian heartlands. Of these, the term that has the longest history in international usage is 'Turkestan/Turkistan'. A word of Persian origin meaning 'land of the Turks', the earliest known occurrence is in a seventh-century Armenian source.[3] In the 9th and 10th centuries it was used by Persian and Arab geographers to refer to the region north of the Syr Darya, outside the orbit of Islam. Centuries later, European travellers and scholars applied the term to the broad sweep of Islamic, mainly Turkic-speaking, lands to the east of the Caspian Sea. More narrowly, it specified a province in north-west Afghanistan.[4]

In the second half of the 19th century, 'Turkestan' acquired a political connotation when the term was incorporated into the official designation of the Tsarist colonial administration in the newly conquered Asian territories, hence the 'Governorate General of Turkestan'. Less formally, this region was known as 'Western' or 'Russian Turkestan', in contrast to 'Eastern' or 'Chinese Turkestan' (i.e. Xinjiang and adjacent lands under Chinese rule). After the Bolshevik Revolution, 'Turkestan' was retained as the name of an Autonomous Soviet Socialist Republic. This lasted a mere six years (1918–24), until the creation of new administrative divisions, based on the national delimitation of Soviet Central Asia (see **Political Geography** below). Thereafter, use of the term 'Turkestan' in a contemporary political sense (as opposed to a historical connotation) was abolished in the Soviet Union;[5] emigrants and some Western writers, however, continued to employ the term in this meaning.[6] Today it still has a political meaning for some groups, particularly those seeking independence for Xinjiang and other regions under Chinese rule.

The earliest term to be used in English for the central Eurasian steppe was 'Tartarie' (i.e. 'land of the Tatar-Mongols'), first recorded in the 14th century.[7] In the 20th century it was still in use, though more as a deliberate archaism than as a current term. By this time the region (understood to lie somewhere at 'the heart of Asia', but of vague size and configuration) had begun to attract not only academic, but also considerable political and military interest. Consequently, there was an increase in writings of various types on the area – journalistic articles, travel books, scholarly studies, official intelligence reports and the like – and also a proliferation of the terms used (more or less synonymously) to refer to it. These included, apart from 'Turkestan', such designations as 'High Asia', 'High Tartary', 'Chinese Tartary', 'Inmost Asia', 'Inner Asia', the 'Heart of Asia', and 'Central Asia'. The term 'Transoxiana' (also rendered as 'Transoxania'), meaning 'across the Oxus' (the Oxus being the classical European name for the Amu Darya), which corresponded to the medieval Arabic appellation *Mā warā' al-Nahr* ('that which lies beyond the river'), was also quite widely used, though generally only with reference to the area located between the Amu Darya and the Syr Darya (roughly corresponding to the territory of modern Uzbekistan).

Today, the terms that are most commonly used by foreign commentators are 'Inner Asia' and 'Central Asia'. Many writers use them synonymously, but there are others who argue that a nomenclatural distinction should be made between regions that, though not clearly separated from each other, are yet in several respects markedly different.[8] However, during the last decade, the term 'Central Asia' has been gaining greater currency than 'Inner Asia'. This was at first mainly in relation to contemporary affairs, but now appears to be a more general shift. It reflects a trend, emanating both from within and without the region, towards the simplification and consolidation of the concept of 'Central Asia' as a discrete geographic entity and system.

Boundaries

The lack of clarity in the terminology relating to the Eurasian heartlands is matched by even greater uncertainty regarding the spatial limits of the region. Thus, while it is now more or less generally agreed that 'Central Asia' is an appropriate name for the region, there is still no consensus as to its territorial limits. In other words, where does Central Asia begin and where does it end? The problem of setting bounds to a notional area is of course not unique to Central Asia. The geographic labels that are conventionally assigned to parts of the globe are always approximate references, and, unless they happen to be clearly demarcated by nature (e.g. islands), or by man (e.g. national frontiers or union boundaries as stipulated in relevant treaties), largely arbitary.[9] Terms such as 'the Middle East', the 'Far East', or even 'Europe' and 'Asia', are from time to time redefined, the concept sometimes widened, sometimes narrowed, depending on a whole host of cultural, political and economic considerations. Nevertheless, in most cases there is a broad and durable enough agreement as to the definition of the core area for it to be possible to use such labels as a readily understood shorthand in popular as well as scholarly contexts. The chief reason for this is that the contours of a sufficient number of conceptual maps (religious, ethnic, linguistic, political etc.) are perceived (rightly or wrongly) to be roughly congruent. This creates an impression of an integrated group identity, which in turn functions as the basis for the mapping out of the spatial limits of the region in question.

This is not so in Central Asia. A place of huge cultural and ethnic complexity, there is very little convergence between the different conceptual geographies. Moreover, the boundaries of these geographies have often changed dramatically and in some cases are still in flux. It can best be described as a four-dimensional kaleidoscope in which social, economic, political and cultural elements continually collide, overlap, merge, fragment and reform. Such an acute degree of fluidity renders it almost impossible to establish an analytical framework that has validity for anything beyond a very limited chronological span. Indeed, it has been regarded by some scholars as one of the main obstacles to creating a coherent history of the region and hence to 'make sense of events there'.[10]

Several attempts have been made to overcome this perceived difficulty. Various sets of criteria have been proposed as the basis for the demarcation of the region (or regions). The first to address this task was the German geographer and explorer Alexander von Humboldt, who located 'Central Asia' in the steppe-desert region, between the Caspian Sea and eastern Mongolia. Other nineteenth-century geographers, among them Ferdinand von Richthofen and Ivan Mushketov, disagreed with this formulation since it did not pay enough attention to the study of physical characteristics, particularly hydrology.[11] They put forward their own proposals, which, however, also failed to win general acceptance. In the early 20th century strategic and geopolitical considerations began to enter the debate.[12] Later, emphasis began to be placed on historical and cultural criteria.[13] Most recently, as mentioned above, political and economic factors have emerged as potential boundary markers. However, there is not as yet any general agreement in popular or in academic circles as to the spatial dimensions of Central Asia. Hence, whenever the term is used, it is still necessary to define the region under discussion if misapprehensions are to be avoided.

Physical Geography

There are no obvious physical features that could serve as boundary markers to delimit a specific Central Asian region within the broader central Eurasian landmass. For this reason, it is sometimes argued that rather than drawing artificial borders, the entire continental heartland should be regarded as constituting 'Central Asia'. In this section an overview will be given of the characteristics of this larger region.

Central Asia, in this wide sense, takes the form of a broad band that stretches latitudinally across the breadth of Europe and Asia. Longitudinally, the band is narrower at its outer margins than in the centre.[14] Characteristic features include rolling grasslands in the northern tier, deserts in the centre and a crest of high mountains in the far south. The climate is extreme, especially in the central section, with great seasonal and diurnal variations in temperature. Water is scarce in many areas; in the east, most of the rivers drain inland. Precipitation is low and evaporation high, creating a harsh, arid environment. This is often exacerbated by biting winds, blizzards and dust storms. Population densities are generally very low, except along river courses and in the vicinity of oases. The region falls into a number of distinct zones. However, there are no sharp divisions between them: they merge into each other through transitional belts.

In the far north, above the limits of tree growth, lies the tundra. The winters here are long and severe, the temperatures subarctic. The ground is extensively covered with permafrost. During the brief summer the top layer thaws, but the ice below impedes drainage; as a result, great waterlogged bogs of mud are created. The vegetation consists mainly of mosses, lichens, woody shrubs and berry-bearing bushes. The reindeer is the mainstay of the economy, providing transportation, food, clothing and furnishings. The tundra is very sparsely populated. The local population, amongst whom there are Turkic, Mongol and Samoyedic groups, have traditionally lived by hunting and trapping.

The forest zone (taiga) stretches from the Baltic Sea to the Sea of Okhotsk. It consists mainly of coniferous trees such as pine, fir and spruce. Along parts of its southern margins, in the Volga region in the west and Manchuria in the east, there are areas of deciduous, or mixed coniferous and deciduous trees. The central section of the taiga is dissected north to south by the Ural mountains, the conventional boundary between Europe and Asia. Compared with the mountains of the south, these are of relatively low elevation and are easy to traverse. The climate in the forest zone is still extreme, though not as rigourous as in the tundra. Here, too, the reindeer predominates. Most parts of the taiga are thinly populated; in the west, the indigenous peoples are mainly Turkic and Finno-Ugric, in the east Mongol and Tungusic.

The forests merge into the steppe zone, a broad expanse of grassland that stretches from the Hungarian and Ukrainian plains, across southern Siberia and the Kazakh steppe, to the Mongolian plateau and Manchuria. This is the natural habitat of the wild horse. The Altai mountains, running diagonally from north-west to south-east across the centre, have traditionally served as a boundary between different states, peoples and cultures. Beyond, further to the east, lies another, lower range of mountains, the Big Khingan; historically, they posed far less of a barrier to population movements and cultural interchange than did the Altai. The vegetation in the steppe zone provides good fodder for most of the year. In the north there are tall meadow grasses; further south, this yields to tussock grass, then to sandy soil with clumps of short, scrubby grass cover. The climate is continental, with summer temperatures of 20°–25°C, winter temperatures of minus 20°–25°C. Most parts of this zone are well watered, with an abundance of lakes and rivers. In the west, the main river systems include the Danube, Dnieper and Don (which flow into the Black Sea) and Volga and Ural (which flow into the Caspian Sea); in the centre, there are the Irtysh-Ob, Yenisei and Lena, all of which flow northwards into the Arctic Ocean. In the east, the rivers drain inland, except for those which flow along the periphery, such as the Amur which drains into the Sea of Okhotsk. The majority of the population in this zone have traditionally been nomadic pastoralists.

To the south of the steppe lies a belt of semi-desert, and beyond that hot deserts (summer day temperatures often reach some 50°C). Here the camel is used alongside the horse. The semi-desert tilts upwards from the low-lying northern shores of the Caspian Sea, across the Turgai tableland of central Kazakhstan, to the Mongolian plateau. The deserts include the Kyzyl Kum ('Red Sands') and the Kara Kum ('Black Sands') in the west, and the Lop Nor, Taklamakan and Gobi in the east. In the centre, zigzagging diagonally across the semi-desert and desert zone, rise the Tien Shan mountains. Historically, they constituted the partition between western and eastern Turkestan. There are passes that lead across these mountains, but they are frequently hazardous. This, coupled with the rigours of the eastern deserts (the Taklamakan is reputed to be the hottest, most desolate and desiccated place on earth), constituted a considerable impediment to the free flow of traffic between east and west. The region has always been thinly populated. Water is scarce and the vegetation sparse. The nomadic animal herders who inhabit these zones follow regular patterns of transhumance, covering a thousand or more kilometres in the course of their annual migration.

The desert zone is transected by a chain of oases and river beds. The most fertile part lies to the west of the Tien Shan, between the Amu Darya and the Syr Darya (i.e. Transoxiana). In the east, the oases are more isolated and further apart, the rivers shorter and less powerful; the main water course here is the meandering Tarim Darya. At the eastern end of the Tarim basin opens the Gansu 'corridor', from time immemorial the principal channel for east-west, west-east contacts and movements of population. Conditions in the oasis-river belt favoured the development of settled communities. The soil is potentially highly productive, but requires irrigation. Agriculture has been practised here for over two thousand years.

The southern margin of Central Asia, from the Caspian Sea to central China and the Burmese border, is dominated by towering mountain ranges and high, wind-swept plateaux. The climate is harsh and extreme, with recorded winter temperatures of minus 60°C in some places. Vegetation is meagre; the upper flanks of the mountains are covered by snow all the year round. This is the domain of the yak. Most of the local population are engaged in small-scale subsistence farming, supplemented by simple arts and crafts (e.g. felt-making, knitting, wood-carving). There is also some nomadic animal husbandry. The highest part of the mountain zone – the 'roof of the world' – is at the eastern end, where the Kunlun Shan and the Himalayan ranges frame the Tibetan plateau (average altitude over 5,000m.). Several of the great rivers of south and south-east Asia rise here, including the Indus and Sutlej (which flow into the Arabian Sea), the Brahmaputra and Salween (which flow into the Bay of Bengal), the Mekong (which flows into the South China Sea), and the Yangze and Yellow River (which flow into the East China Sea and the Yellow Sea respectively).

The southern foothills of the Himalayas are fringed by a band of small states and principalities – Ladakh, Nepal, Bhutan and Sikkim.

To the north-west of the Himalayas rise the Pamir, Hindu Kush and Karakorum ranges, which fan out into Kashmir, the North-West Frontier Provinces of Pakistan, Tajikistan and Afghanistan. Further south, beyond this massive mountain barrier, are located, to the east, the Indus valley, and to the west, the uplands of Afghanistan and Baluchistan. Routes across the mountains are few and mostly treacherous; in the winter they are snow-bound. From ancient times the main route through the Hindu Kush has been along the Kabul valley to the Khyber Pass; however, other ways have also been used and some are now being redeveloped, such as the Karakorum Highway via Gilgit. Westwards, stretching from this tangled knot of high mountains to the Caspian Sea, there are the Elbruz and Kopet Dag ranges. To the south lies the Iranian plateau, to the north the Kara Kum desert. Although the peaks of these mountains are lower than those of the central ranges, nevertheless, they are barren and steep, a forbidding obstacle to travellers. Historically, the main north-south route ran from Merv (modern Mary, Turkmenistan) across the Tedzhen river to the cities of northern Iran.[15]

Historical and Ethnic Geography

Just as there are no natural physical boundaries that mark off a Central Asian region from the larger mass of the Eurasian continent, so, too, there are no obvious historical features that provide a framework within which to locate the study of Central Asia. Whatever limits are chosen are likely to be based on subjective considerations, as, for example, the demands of ideologised national historiographies; or particular disciplinary interests, such as the study of philological contacts; or snapshots of specific periods during which, for a longer or shorter span of time, patterns of linkages emerged which imposed, for a while at least, certain recognisable boundaries (e.g. 'Islamic' Central Asia, 'Kushan' Central Asia, 'Soviet' Central Asia).

These 'demarcations' do not of course offer a definitive solution to the question 'what are the limits of Central Asia?', but they are convenient tools because, within their own terms of reference, they establish a set of bearings for the study of some portion of this huge, amorphous territory. Likewise, in this and subsequent sections, as a practical means of limiting the field the primary focus will be the geographic area that is covered by the CoDoCA project. Thus,

11

hereafter 'Central Asia' is to be understood as referring to the area that is bounded by the Caspian Sea in the west and the Gobi desert in the east, by the steppe in the north and the Himalayan and Hindu Kush mountain ranges in the south. A justification for this definition is suggested in the final section of this chapter (see **Environmental Awareness and Regional Identity** below).

The time span that is covered in this chapter may seem inordinately long, but without this broad perspective it is difficult to appreciate the complexity of the historical legacy. Moreover, today, at popular, academic and official government levels, elements from modern as well as very ancient history are being used as building blocks in the construction of new national and regional identities. Consequently, it is not possible to select any one period as 'representative' or 'definitive': rather, it is the whole spectrum of historical development that is the defining force. Unfortunately, there are very few works in western European languages that treat this *longue durée*, hence the need for this lengthy introduction.[16]

The history of Central Asia since earliest times has been marked by large-scale migrations of people. These have been mostly from east to west, but there have been significant flows in the opposite direction, and also some movement southwards. These shifts often resulted in the emergence of state formations that were highly fluid – loose tribal confederations with unstable boundaries. Some of these groupings lasted no more that a few decades, others, in one form or another, for several centuries. Most were destroyed by new waves of invaders. All were multi-ethnic, with a culture that was usually syncretic in nature.They encompassed huge areas, many times the size of most modern European countries. The degree of geographic mobility, both of individuals and of large armies, was extraordinary even by present-day standards. This long-distance travel facilitated the development of cultural, economic and technological exchanges over a wide area. It also brought about political and military contacts, sometimes amicable, sometimes hostile, between powers that, from today's perspective seem very far apart indeed, such as, for example, the Arabs and the Tibetans, or the Turks of Mongolia and the Byzantine court.

A crucial regional dynamic was the long interaction between the nomad states of the steppe-desert zone and the centralised sedentary empires around the periphery. The main pole of attraction in the south-west was Iran; in the south-centre, India; in the east, China; and in the north (much later, but then very powerfully), Russia. Even when these empires did not physically control territory in Central Asia they exerted great cultural and economic influence over contiguous areas; often, too, though less directly, over more distant regions. It was by no means a one-way process, however. The influence of the nomads on the surrounding sedentary civilisations was not inconsiderable. This is not surprising given that nomad hordes periodically conquered large swathes of the settled lands. Moreover, even in times of peace, they were linked by military alliances and trade to their sedentary neighbours. Contacts such as these helped to spread elements of nomad culture well beyond the confines of the steppe-desert zone.

3rd to late 1st millenium BC: Empires of the Steppe and the Sown

From distant prehistory Central Asia has been a place of encounter betweeen peoples of different races and cultures. The earliest contacts were between Europoids (proto-Indo-Europeans)and Mongoloids (including proto-Mongols, proto-Turks, proto-Tibetans and proto-Tungus). Archaeological remains dating back to c. 3,000 BC show that the former had settlements on the steppe extending from the Urals to the Tarim basin (eastern Turkestan), while the latter inhabited the forests and the more easterly uplands. The first historically identifiable peoples are the mounted Iranian nomads who dominated the western steppe in the 1st millenium BC. They included the Scythians, whose original habitat (c. 600 BC) was the Pontic steppe north of the Black Sea; the Sarmatians, further to the east; and the Sakas, located to the north-east of Iran, up to the foothills of the Pamirs. Some of the Sakas later migrated eastwards to the Tarim basin (where they adopted a settled way of life and established the beginnings of an urbanised culture), while other groups moved south through the Pamirs into India.[17]

The first great sedentary power to incorporate a significant part of Central Asia within its boundaries was the Achaemenid empire of Iran. The state already included Babylonia, Syria, Palestine, Phoenicia and Asia Minor when, in the mid-6th century BC, it extended its rule northwards into Central Asia. Several new provinces (satrapies) were established on what is today the territory of Afghanistan, Turkmenistan and Uzbekistan. These included Bactria (south of the Amu Darya), Chorasmia (south of the Aral Sea), Margiana (centred on the Merv oasis) and Sogdiana (Zerafshan valley). It is very likely that the lands across the Amu Darya were never more than nominally under

Achaemenian rule. Even so, tenuous though the link may have been, it enabled this part of Central Asia to be brought within the reach of cultural and economic influences from Iran. During this period sedentary Iranian groups began to settle in the region, especially in Chorasmia. They introduced irrigated agriculture and established the beginnings of an urban culture.

In east Central Asia mounted nomadism appeared somewhat later than in the west. By about 500 BC, however, it was fully developed and this paved the way for the rise of the first of the many great nomad empires of the eastern steppe, that of the Hsiung-nu. There is no scholarly consensus as to the origins of this people (or group of peoples). Some authorities have suggested that they were Turkic-speakers, and possibly the forebears of the Huns who, many centuries later, under the leadership of Attila, were to ravage central Europe.[18] By the late 3rd century BC the Hsiung-nu were the dominant regional power. Their sudden and rapid expansion had a number of far-reaching consequences. One was that on their eastern flank they began increasingly to encroach on Chinese territory. A long power struggle ensued which eventually (60 BC) resulted in the rout of the Hsiung-nu. In pursuit of this war, the Chinese made incursions into the Tarim basin: this culminated in the eastern part of Central Asia coming under Chinese rule.

Another major consequence of the rise of the Hsiung-nu was that it caused the displacement of neighbouring peoples, setting in motion a chain reaction of population movements. One of the displaced groups was the Tung-hu, a Tungusic people, most of whom subsequently moved eastwards to the territory of modern Manchuria. Another group (or confederation of groups) was a people who are known in Chinese sources as Yüeh-Chih, in Greek and Latin sources as Tocharians. Little is known of their origins other than that they spoke an Indo-European language (two forms existed, classfied today as Tocharian A and Tocharian B) which was closer to Hittite than to the languages of the geographically closer Iranian steppe peoples. The Yüeh-Chih (i.e. Tocharians), expelled from their original habitat in the eastern steppe, moved westwards through Dzungaria to the upper reaches of the Ili, Chu and Naryn rivers. Subsequently, some of them – generally known as the Great Yüeh-Chih – migrated to the Amu Darya (c. 175 BC). They later crossed the river into Bactria, where they encountered Greek colonies.

4th century BC to 3rd century AD: Greeks, Parthians and Kushans

Greek rule in southern Central Asia was established by Alexander the Great who, in the second half of the 4th century BC overthrew the Achaemenid dynasty and took possession of their empire. His victorious campaign carried him through the Ferghana valley (modern Uzbekistan) up to the foothills of the Pamirs. Here he encountered fierce resistance and did not succeed in fully subduing Transoxiana. Further south, however, his impact was far greater. Although Alexander remained in this part of Central Asia for scarcely two years (328–27 BC), he left a lasting imprint on the region. He founded, or restored, a large number of cities, among them (using present-day place names) Herat, Kandahar, Ghazni (all in modern Afghanistan), Mary (modern Turkmenistan) and Khojand (modern Tajikistan). He settled them with thousands of Greek colonisers, including soldiers, administrators, traders and craftsmen (see **Cultural Geography** and **Economic Geography** below).

Soon after Alexander the Great's death (323 BC) Greek rule in southern Central Asia-northern India began to atomise. There was a power struggle among his successors which resulted in the loss of control over the Central Asian provinces. Greek dominance was re-established in some parts c. 305 BC, under the Seleucids, but by this time the southern margin (Kandahar-Kabul) had fallen under the influence of the Mauryan empire of India and the Greeks had little option but to accept the suzerainty of the latter here.

In the middle of the 3rd century BC, two new states emerged along the south-western margins of Central Asia, on the territory of Alexander's empire. The larger of these was Parthia, founded by Iranian nomads. Mounted archers of legendary prowess, they established an empire that was eventually to stretch from north of the Kopet Dag (modern Turkmenistan) down to the Persian Gulf, and from the Euphrates across to the Hindu Kush. In 53 BC they gained their first major victory over the Roman army. By the 1st century AD, they were one of the four greatest powers in the world, on a par with Rome, China and India. One of their capitals was at Nisa, near Ashghabat, the capital of present-day Turkmenistan. As part of the Parthian empire, even though on the periphery, this corner of south-western Central Asia was again tied into the cultural, economic and political world of the Middle East.

15

The other state that took shape at this same period, also to the south of the Amu Darya, was Graeco-Bactria. Much smaller in size than Parthia, it encompassed part of the Zerafshan valley (modern Uzbekistan), most of the Hindu Kush (modern Tajikistan and Afghanistan) and the upper Indus valley. A Greek dynasty continued to rule here until the 2nd century BC, when it was dislodged by the incoming Yüeh-Chih. Further south, smaller Indo-Greek principalities survived for another century or so, until they, too, were overthrown by Central Asian invaders (among them groups of Sakas, who subsequently founded kingdoms on the territory of what is today Afghanistan and northern India).

The state established by the Yüeh-Chih (possibly with the assistance of other groups) on the territory of Graeco-Bactria has come to be known as the Kushan empire. Many aspects of the history of this state are still unclear, or a matter of dispute amongst scholars.[19] The state reached its zenith in the first half of the 2nd century AD. At this period its lands covered an immense triangle, the apex of which extended as far north as the Syr Darya, almost up to Tashkent (modern Uzbekistan). Its western border stretched southwards through Baluchistan to the Arabian Sea, its eastern border across the Karakorum and Pamir ranges to the Ganges (approximately to the level of Benares). This configuration linked the central-southern region of Central Asia to the Indian subcontinent and consequently opened the way for the formation of new cultural and economic networks. This added a fresh dimension to the development of southern Central Asia, especially in the spheres of art and religion.

Thus, to summarise the situation in the 1st and 2nd centuries AD, the southern margin of Central Asia either formed part of, or was under the influence of, three great empires: in the west, the Parthians; in the centre, the Kushans; and in the east the Han Chinese. This was a time of relative stability. Relations between the three powers varied, but were generally cordial rather than hostile. The Kushans – not surprisingly, given their geographic location – played a pivotal role in facilitating transcontinental trade, also in shaping transcontinental diplomatic relations. This sometimes included offering assistance to the Chinese in their campaigns to subdue the Tarim basin. The influence of these empires on the periphery did not, however, penetrate far into the interior of Central Asia. Further north, in the steppe zone, tribal leaders held sway, maintaining a fluctuating balance of power amongst themselves.

3rd century AD to 8th century AD: White Huns, Turks and Arabs

By the 3rd century AD, the situation along the southern rim had changed. In the west, a new power had emerged, the Sassanids of Iran, who conquered the lands of the Parthians and much of that of the Kushans. Thus, once again, the south-west and south-central regions of Central Asia (modern Turkmenistan and Uzbekistan) were incorporated into an Iranian empire.the Sassanids, unlike the Parthians, came from a sedentary culture. In many ways, the state they created represented a continuation of the imperial Achaemenid tradition. During this period there was a revival and expansion of cultural, economic and political influences emanating from Iran in the region. In the east, meanwhile, China's hold on the Tarim basin had weakened to the point where it was scarcely even nominal; local rulers established virtually autonomous fiefdoms in the environs of the main oases.

The 5th century ushered in yet another period of upheaval. It saw the start of a sequence of invasions, accompanied by mass shifts of population, that was to continue for the next three centuries, up to the mid-8th century. These movements set in motion a process of change that eventually was to result in a radical transformation of the ethnic and cultural map of almost all of Central Asia. The first onslaught was the incursion of the Hephthalites (also known as the 'White Huns'), who swept across southern Central Asia in the second half of the 5th century. Their ethnic origin and early history is not clear: it has been suggested that they might be related to the Avars, or possibly be descendants of a section of the Yüeh-Chih.[20] By the early 6th century they controlled most of the Tarim basin, the Ferghana valley and territory further to the west, to the borders of Sassanian Iran.

The second great movement was that of the Turks. Their original homeland appears to have been in Mongolia, but it is possible that some Turkic groups were also located in the northen steppe and forest zone (especially in the vicinity of Lake Balkhash and the Volga). This remains a matter for speculation, however, since the first historically attested appearance of a Turkic people dates from the mid-6th century, from the territory of the Orkhon basin (modern Mongolia). They were at this time under pressure from the Chinese further to the east. Turkic tribes began to fan out from the central steppe zone, dispelling the Hephthalites and other regional nomadic confederations. In an extraordinarily short space of time they established control, albeit fluctuatingly, over a vast domain that, at times,

17

stretched from Manchuria to the Aral Sea. Their lands now included not only the steppe, but also parts of the cultivated oasis belt in the south, where the predominant population was Iranian. This empire, generally known as the Turkic Khaghanate, was powerful and important enough to enter into diplomatic negotiations with Byzantium in AD 568. It also maintained relations with Iran and China. However, it was a fragile entity, with little internal cohesion. Before the end of the century the Khaghanate had split into an eastern and a western wing. Both came under Chinese domination for a while, but subsequently regained their independence and survived till the mid-8th century.

In the west, meanwhile, Arab troops were making rapid inroads into Central Asia.[21] Having established control of Khorasan (north-eastern Iran) in the second half of the 7th century, they advanced northwards across the Amu Darya, into the Ferghana valley (modern Uzbekistan, Tajikistan and Kyrgyzstan), in following century. Here they met with substantial resistance from the local rulers, who turned to China for assistance. This invitation to intervene in the affairs of south-western Central Asia was not unwelcome to the Chinese, who were still trying to undermine the Turkic Khaghanate. Consequently, they mustered a large army and joined battle with the Arabs on the Talas river (modern Kyrgyzstan) in AD 751. The supply lines of both parties were over-extended and they had to rely heavily on local allies of dubious loyalty. In the event, the result was a resounding triumph for the Arabs and a humiliating defeat for the Chinese.

Yet the Arabs had little interest in pursuing their campaign eastwards across the Tien Shan. Instead of undertaking new advances, they concentrated on consolidating their position in south-western Central Asia (Transoxiana). Nevertheless, as a precautionary measure, they entered into a strategic alliance with the Tibetans, who, further to the east, were also engaged in a struggle against the Chinese. This was the period when, after centuries of disunity, the Tibetans had achieved some degree of consolidation under a single leader. This enabled them to establish a sizeable empire of their own, extending at its height from the Tarim basin in the north to the Gangetic plain, almost to the Bay of Bengal, in the south. These expansionist tendencies alarmed the Arabs, who terminated their pact with the Tibetans in AD 789. Tibetan power began to wane shortly thereafter and by the mid-9th century they were no longer a significant regional force.

8th century AD to 12th century AD: Islamic states

The Arab conquest of Transoxiana fixed south-west Central Asia firmly within the orbit of the Caliphate. Although geographically on the periphery, this region now became fully integrated into the world of Islam. As the local population converted to the new religion they acquired not simply a faith, but became a part of the same intellectual space, participated in the same economic system, and importantly, came to share the same system of values as Muslims in the Middle East. The Islamicisation of the sedentary population of the oasis belt occurred within a relatively short period of time, at most a century or two. It took far longer (some would claim at least a millenium) before Islam encompassed all of the western steppe; the conversion of the eastern steppe petered out at the borders of Mongolia. In the north, it scarcely penetrated the forest zone (see **Cultural Geography** below).

The physical presence of the Arabs in Central Asia was numerically rather small, and of somewhat short duration. Within a generation or so they had begun to intermarry with the local population (mixed Iranian and Turkic by this time) and gradually lost a separate ethnic identity. By the early 9th century, local governors were already beginning to assert their autonomy. In AD 875 the Samanids, a Persian dynasty from Khorasan, made Bukhara their seat and established an independent state that extended from the Ferghana valley south across the Amu Darya to the foothills of the Hindu Kush. This was the first indigenous independent Muslim state in Central Asia.

In the east, during this same period, new influxes of Turkic peoples as yet unconverted to Islam continued to stream out of the Mongolian plateau. These included the Uighurs, who founded a vast semi-sedentary empire on the lands of the former eastern Turkic Khaghanate (i.e. centred on the territory of western Mongolia) in AD 745. They adopted Manichaeism as their state religion. They were overthrown by another Turkic people, the Yenisei Kyrgyz, c. AD 840. However, Uighur statelets survived for sometime longer in Gansu and the Turfan oasis.

The first Turkic Muslim (or at least partially Muslim) dynasty in Central Asia was that of the Karakhanids. It was formed on the basis of a tribal confederation that, in the early 10th century, extended its rule in a wide band across central Turkestan, spanning the Tien Shan to east and west. The main cities of the Karakhanids included Balasaghun (near Lake Issyk Kul, modern Kyrgyzstan) and Kashgar (modern Xinjiang). In AD 999 the Karakhanids captured Bukhara, ousting the Samanids and formally establishing Turkic dominance of the region. Further to the west there were already a number of other Turkic states, offshoots of earlier migrations. These included the Oghuz (forebears of the Seljuks and Turkmens) in the Aral basin; the Khazars to the west of the Volga (who were at the height of their very considerable power in the 8th century, when, too, the ruling elite and some of the population converted to Judaism); and the Pechenegs, Cumans and other nomadic peoples of the southern Russian steppe.

In the early 12th century a new wave of Central Asian nomads invaded Central Asia. These were the Kitans (known in western Central Asia as the Kara Khitai), a Manchurian people who originated in the Jehol province in the basin of the Liao river. They surged across the steppe, overthrowing the Karakhanids and likewise the Oghuz Seljuks. By 1124 they were masters of an immense semi-nomadic state that stretched from the borders of China to the Aral Sea. They made Balasaghun, a former Karakhanid stronghold, their capital. The Kitans succeeded in retaining control of the region for almost a century, until they, in turn, were overthrown by a fresh invasion from the east – that of the Mongols. The Kitans did not recover from this defeat. They were assimilated by the incomers and by the local population and by the 14th century had ceased to exist as an identifiable, separate people.

13th century AD to 16th century AD: Mongols and Timurids

The Mongols were the driving force of the last great eruption of nomad power. The unification of the Mongols and neighbouring tribes (amongst them Turkic and Tungusic groups) occurred under the leadership of Genghiz Khan in the first decade of the 13th century. Thereafter, having consolidated the social and military reorganisation of society, he led his mounted troops on a series of campaigns of awe-inspiring scope and audacity. In 1215 they captured Peking; a few years later, they were masters of Turkestan and its ancient cities such as Bukhara, Samarkand, Tashkent and Khojand.

Their further conquests brought most of the known world under Mongol rule, including large parts of the Middle East, Transcaucasia, the Far East, northern India and Afghanistan, south-east Asia, and eastern and central Europe.[22] For the first and only time in its long history, Central Asia, in the widest geographic interpretation of the term, was united under a single ruler. However, before his death (AD 1227) Genghiz Khan had already divided his empire into vassal states, each headed by one of his sons. Initially subject to the overlordship of the supreme Khan, these were eventually to become fully independent entities. The lands to the west were assigned to Genghiz Khan's eldest son Juchi; this state, which later included much of Russia, came to be known as the 'Golden Horde'.[23] Most of Turkestan (from the Aral Sea to the Altai mountains) came under Genghiz Khan's second son, Chagatai. The central heartlands came under the youngest son, Tuli. One of Tuli's sons, Hulagu, founded the dynasty of the Ilkhans of Persia; another, Kubilai, the Yuan dynasty of China.

In the first years after Genghiz Khan's death the capital of the entire empire continued to be Karakorum (Kharkhorin), on the Orkhon river. When Kubilai was elected Great Khan, however, he moved his seat to Peking. Thereafter Mongolia itself gradually became an impoverished backwater. The Mongol element in each wing of the empire became increasingly influenced by the local culture. In the lands of Juchi and Chagatai (i.e. from Turkestan to the Volga) they became Turkicised and Islamicised; in the Ilkhanate, Persianised and Islamicised; in China, Sinicised. In Mongolia itself, Tibetan culture and religious practices became predominant.

Another mighty Central Asian empire was founded at the end of the 14th century by Timur, or as he is more usually known in the West, Tamerlane. In character, this state was a rich fusion of nomad and sedentary traditions.[24] Tamerlane himself exemplified this mingling of cultures: a descendant of the Mongol Barlas tribe, he was Turkic-speaking and Muslim. He excelled in the skills and tactics of nomad warfare, but deeply admired and cultivated the legacy of Persian culture and statecraft that still flourished in the cities of Transoxiana. He established his capital at Samarkand and made it a leading centre for the development of science, art and architecture. His conquests, though not as extensive as those of Genghiz Khan, nevertheless encompassed a huge swathe of Khorasan, Transcaucasia, northern India, parts of the Middle East and Asia minor, and much of Russia. He died in 1405, on a campaign to China.

After Tamerlane's death, the Timurid dynasty, and likewise the empire, fragmented even more rapidly than that of Genghiz Khan. Ever smaller tribal and territorial entities acquired virtual independence. The leaders were (or at least claimed to be) of Mongol descent, of the house of Genghiz Khan, but the overwhelming bulk of the population was Turkic. The Turkic element was strengthened as some of the tribes that had originally moved westwards with the armies of the Golden Horde now began to flow back towards the east. In

Transoxiana they adopted a sedentary, or semi-sedentary way of life, and merged with previous Turkic settlers in the region; these had already assimilated in some degree (primarily in the urban areas) to the ancient Persian culture of the region. These different strands gradually coalesced to form the Uzbek people. Similarly in the steppe, there occurred a fusion of various tribal elements (mainly Turkic, but also Mongol and possibly some earlier ethnic strata) and this in time gave rise to the formation of the modern peoples of the region, such as the Kazakhs, Kyrgyz and Uighurs.

Tamerlane was not the last of the great Central Asian conquerors. Almost a century later, one of his descendants was to win an empire that was almost as large and to found a dynasty that was to remain in power considerably longer. However, this was not in Central Asia: Babur, heir to a minor principality in the Ferghana valley, was ousted from his patrimonial lands by powerful Uzbek tribes in the early 16th century. He sought refuge across the Amu Darya, capturing Kabul and making it his capital in 1504. In 1526 he defeated the Sultan of Delhi at the battle of Panipat and thereby laid the foundations of the Moghul (Arabic/Persian form of 'Mongol') empire in India. He had hoped to reconquer Transoxiana and to establish his seat, like Tamerlane, at Samarkand. In this, however, he failed; Central Asia remained outside the bounds of the Moghul sphere.

Ethnic distribution

The 16th century marked the end of the main shifts of population in Central Asia. Migration to and from the region continued (especially in the 20th century), but it was on a smaller scale than previously. Thus, after some two millenia of constant movement of peoples the ethnic map of the region finally stabilised, assuming, in broad outline, the pattern which still prevails today. The greatest transformation which took place during the period of pre- and early history was the change in the geographic distribution of Europoids and Mongoloids, and more specifically within those two groups, of the Iranian and Turkic peoples respectively. To recapitulate briefly, the Iranians dominated the steppe and oases from the Caspian Sea to the western edge of the Mongolian plateau up till the 6th century AD. Thereafter, they were expelled or submerged by successive waves of insurgent Turks from the east. The boundary between the Turkic and Iranian ethnic and linguistic spheres in Central Asia came to be, in approximate terms, the Kopet Dag and Hindu Kush mountain ranges and the Amu Darya river, with the Turkic peoples located mainly to the north of this line and the Iranian peoples to the south. The chief exceptions to this division are, amongst the Iranians, the Tajiks of Tajikistan and of the Bukhara-Samarkand area of Uzbekistan; amongst the Turks, some Turkmen tribes and a few nomadic groups that are located in northen Iran.

The Mongols did not leave much of a separate ethnic presence far outside their original homeland on the Mongolian plateau. Today they are mainly located in Mongolia (formerly 'Outer'), Inner Mongolia (China), and Buryatia (southern Siberia, Russian Federation); there are also Mongols in Kalmykia (lower Volga river, Russian Federation), the descendants of seventeenth-century migrants; and Mongoloid Hazaras in Afghanistan (probably descended from Genghiz Khan's troops).[25] The Turkic tribes who originally inhabited the Mongolian plateau have mostly either migrated out of the region or have been absorbed by the neighbouring peoples. The approximate boundary between the Turkic and Mongol peoples of the steppe zone is now the Altai range of mountains. Further north, however, in the taiga and tundra, almost as far east as the Bering Straits, Turkic groups (e.g. Yakuts, Tuvinians, Khakass and Dolgans) are to be found, dispersed among Mongol, Tungusic and Samoyedic peoples.

The majority of the Tungus people, who also originated in Mongolia, are now mostly located in Manchuria. Tibetans are spread over an extensive area of southern Central Asia. The high plateau, their historic homeland, is still the main area of Tibetan settlement, but there has been some movement into the adjacent regions of Qinghai, Sichuan, Gansu, Yunnan and Xinjiang, also into the southern foothills of the Himalayas. Some of this dispersal occurred in the 8th century, when the Tibetan empire was at its zenith and held sway over this territory; other migrations took place later, most notably into northern India after the imposition of Communist rule in the second half of the 20th century. Of the other peoples who held sway in Central Asia, none have left distinct ethnic enclaves, although as regards the Greeks and the Arabs, their physiognomic imprint is still very clearly to be discerned in the areas in which they settled during their occupation of the region.[26]

Political Geography

The modern political geography of Central Asia began to take shape in the 17th century. After the disintegration of the empires of Genghiz

23

Khan and Tamerlane the region had dissolved into an array of independent entities – nomad confederations in the steppe and desert, city states in the oasis belt, small fiefdoms in the foothills of the high mountains. The main groupings included the three Kazakh 'hordes' (*zhuz*) – known as the Big, Little and Middle – in the western steppe; the Turkmen tribes in the south; the oasis-based principalities with predominantly settled populations in the south-centre; the Dzungarian Oirats (western branch of the Mongols) who, east of Lake Balkhash, created the last significant Central Asian nomad state; the Khalkas (eastern Mongols), on the territory of present-day Mongolia, who represented the rump of the original Mongol empire; and the theocratic Tibetan state ruled by the Dalai Lama on the Tibetan plateau.

The surrounding centralised empires were either too weak or too preoccupied with events elsewhere to devote much attention to the Eurasian hinterland during this period. However, from the 17th century onwards the situation changed as Russia and China, and later British India, vied for influence in the region. Of the traditional regional players, only Iran remained too feeble to take part in this struggle. Great power rivalry between the other three – the 'Great Game' – intensified in the 19th century. By the turn of the century, the whole of Central Asia was under the domination, directly or indirectly, of one of these states. Despite the upheavals caused by war, revolution and changes of regime that subsequently racked the region, the divisions that were established at this time remained in force, with only a few exceptions, until the collapse of Soviet power and the re-emergence of a number of independent countries in the region at the end of the 20th century.

Chinese expansion

The Chinese expansion into Central Asia (re-establishing control over areas that had at least nominally formed part of Chinese empires in earlier centuries), was preceded by the Manchu rise to power. After unifying under a single leader in 1616, the Manchu tribes began to aspire to the imperial throne of China. In 1644 they finally succeeded in capturing Peking and, having overthrown the incumbent Ming emperor, inaugurated a new, Manchu dynasty, the Qing, which was to last until 1911. The Qing emperors, though they retained some elements of the traditional Manchu way of life, enthusiastically embraced Chinese culture and customs, becoming in the process thoroughly Sinified and true heirs to the Chinese imperial legacy. In the 1680s, under the emperor Kangxi, they embarked on the conquest of the Tarim basin. This brought them into conflict with the Oirats of Dzungaria, who controlled most of eastern Turkestan at this time. It was not until 1758 that the Oirats were at last defeated and the Qing were able to establish Xinjiang, the 'New Dominion', as an administrative entity in this part of Turkestan. During the 17th century the Qing also re-incorporated Mongolia into the Chinese political, administrative and military orbit. The eastern Mongol tribes had allied themselves to the Manchus as early as 1636; their territory was designated 'Inner' Mongolia. However, the Khalkha Mongols to the west did not accept Manchu rule for another 50-odd years. It was the pressure of devastating attacks from the Dzungarian Oirats of eastern Turkestan that finally persuaded them to seek Chinese (i.e. Manchu) protection; in 1691, at the Convention of Dolon Nor, 'Outer' Mongolia became a formal Dependency of the Chinese state.

Tibet in the early 18th century was weak and disunited, with no central political authority. It was also suffering from the depredations of the Oirats, who captured Lhasa in 1717. The Qing chose this period to intervene in Tibetan affairs. The remoteness of the country, however, made it difficult to maintain a Manchu-Chinese bureaucracy here. A boundary was drawn along the watershed of the Yangze and Mekong headwaters, dividing the eastern, 'inner' sector, where a form of protectorate was established, from the western, 'outer' sector, which was left in the hands of the Dalai Lama's government in Lhasa (which was in effect independent, but formally under the suzerainty of the Qing emperor). Once Manchu-Chinese rule had been consolidated in eastern Central Asia (Mongolia, Xinjiang, part of Tibet), substantial immigration commenced into the 'inner' territories, i.e. into those regions that bordered central China. The 'outer' regions remained largely free of immigration, other than that of government officials, traders and soldiers.

In the early 18th century, the Dzungarian Oirats also began to encroach on the Kazakh steppe. Some of the Kazakh tribes at this time looked to the Russian crown for protection (see below), but others sought assistance from the Chinese. After the overthrow of the Oirats (1758), part of the Kazakh Big Horde was incorporated into the Qing empire. The Kazakh lands east of the Tien Shan have remained under Chinese rule up till the present, forming part of Xinjiang. To the north-east of the Kazakh steppe (bordering Mongolia), the Tuvinians, a group of Turkic tribes of southern Siberia, experienced a similar fate to that of the Kazakhs: after decades of domination by the Oirats they were finally incorporated into the Qing empire in the second half of the 18th century.

Russian expansion

Russian expansion eastwards was accomplished in several stages. The first, which took place c. 1580–1644, was the conquest of Siberia from the Urals to the Pacific. It was spearheaded by merchant adventurers and explorers. A series of military bases was established at strategic points across the region; later, administrative units were also put in place. The final thrust of this advance brought the Russians to the Amur river and hence into direct confrontation with the Chinese. Diplomatic relations were instigated (not without considerable difficulty, owing to the reluctance of the Chinese to deal with foreign states), leading to the signing of the Treaty of Nerchinsk in 1689. This specified areas of jurisdiction: both banks of the Amur were to remain under Chinese control, as was Outer Mongolia. The Russo-Chinese Treaty of Kyakhta (1727) clarified aspects of trade and diplomatic contacts between the two parties.

The second stage of the Russian expansion was the move southwards into the Kazakh steppe. This took place 1680–1760. It began when individual Kazakh rulers, menaced by the Dzungarian Oirats from eastern Turkestan, sought help from the the Russians. The latter were beginning to appreciate the strategic importance of this region and were therefore prepared to look favourably on such requests. By the early 19th century, most of the Kazakh tribes had been fully incorporated into the Russian empire; the remainder were under Chinese rule. A line of Russian defensive posts had been established along the north-eastern rim of the Kazakh lands in the course of the 18th century; in the first half of the 19th century this line was steadily pushed further south, reaching the Syr Darya by the middle of the century. A colonial administration was introduced that included, among other provisions, the creation of internal provincial and district boundaries. Some of the Kazakh elite were co-opted into the new system. They became increasingly Russified in their education and way of life and as a result, began to lose contact with traditional clan-tribal Kazakh society.[27]

The next stages of the Russian movement to the east were the conquest of the territory between the Black Sea and the Caspian Sea (1785–1830), and further annexations in the Amur region (1850–60).

This was followed almost immediately by the final stage, namely the conquest of Transoxiana and Transcaspia (approximately corresponding to present-day Uzbekistan and Turkmenistan respectively). This began with the capture of Tashkent in 1865 and was completed with the capture of Merv, the main outpost of Turkmen resistance, in 1884. A colonial administration for the region (the Governorate-General of Turkestan) was created in 1867. It encompassed most of western Turkestan (i.e. the land between the Caspian Sea and the Tien Shan), except for the khanates of Bukhara and Khiva, which retained a nominally independent status as protectorates. The integration of the Central Asian territories into the Russian empire was greatly assisted by the construction of railway lines linking Transoxiana to central Russia; these were of economic as well as strategic importance.[28]

The enlargement of the Tsarist empire caused consternation amongst its neighbours. China was by this time in a weakened state, beset by uprisings and disorders in many parts of the country. Xinjiang was virtually independent, ruled by local warlords; the most powerful of these was Yakub Beg, who in the 1860s-70s tried to establish a separate Eastern Turkestan state. The Russian government, ostensibly in an effort to stabilise the situation, moved army units across the official border with China into the Ili region of Xinjiang in 1871. They took Kuldja, along with a large area of the surrounding territory, and set up a military administration; they insisted that this was a temporary measure. The Qing government was not reassured, however, and continually raised the question of the return of the region to Chinese jurisdiction. It was not until the conclusion of the Treaty of St Petersburg (1881) that this was finally achieved, albeit in return for some concessions to the Russians. Yakub Beg's bid for autonomy had failed and some degree of peace returned to Xinjiang. However, substantial numbers of Muslims (Kazakhs, Uighurs and Dungans), fearful of the return of Chinese domination, moved to Russian Turkestan at this time.

British expansion

In India, where British rule had now replaced Mughal rule, Russia's ever-widening frontiers were also viewed with unease. The British tried to advance their own northern border, occupying Baluchistan and entering into a succession of wars and treaties with Afghanistan. They extended their sphere of influence into the Himalayan mountain states, and tried to counter Russian efforts to penetrate Xinjiang and

Tibet. British efforts to establish a foothold in Xinjiang and Tibet did not meet with great success, but they were able to negotiate a treaty, based on the work of the 1895 Anglo-Russian Boundary Commission, which fixed the northern boundary of Afghanistan and the southern limit to Russian expansion. In part, this ran along the Amu Darya, the river that had so often in the distant past formed the border between competing spheres of influence.

20th century: new political systems, new ideologies

The Manchu dynasty was overthrown in 1911 and replaced by a republic. Outer Mongolia took this opportunity to declare its independence under a theocratic Buddhist sovereign. The leaders of Inner Mongolia did not join this new state. There followed a confused period of constantly changing political affiliations, during which factions in Outer Mongolia looked increasingly to Russia for assistance against continuing Chinese aggression. In Russia, meanwhile, the Tsarist regime was overthrown by the Bolsheviks in 1917. Civil war broke out throughout the empire and Soviet rule was not finally established until 1921–22.

Outer Mongolia was also affected by these events and for a time invaded by White Russians. In 1921 a provisional Mongolian government was set up on Russian-held territory; later that year, with the help of the Red army, they expelled the Chinese from Outer Mongolia. Inner Mongolia remained under Chinese rule.The Mongolian People's Republic (MPR) was proclaimed on the territory of Outer Mongolia in 1924, after the death of the theocratic sovereign who had intermittently held the throne since 1911. The MPR was the first, and many would claim most subservient, Soviet satellite. It was not until 1945 that the country's independence was formally recognised by China and some other members of the international community; it was admitted to the United Nations in 1961. Soviet forces were also instrumental in establishing a satellite state in the Tuva region, which, in 1921, acquired independence as the Tannu-Tuva People's Republic. This was incorporated into the Soviet Union in 1944 and remains part of the Russian Federation today.

Tibet and Xinjiang also established a degree of sovereignty after the collapse of the Manchu dynasty. However, unlike the Mongols of the MPR, they did not have the support of a strong neighbour and failed to consolidate their autonomy. Their international status remained ambiguous. Yet western Tibet did achieve *de facto* independence; it had a credible government, an army and some progress was made in constructing a foreign policy. Britain, despite some equivocation in Whitehall, was on the whole in favour of Tibetan independence and during the inter-war years carried on negotiations with China to secure a settlement of the 'Tibetan question'. However, little progess was made and when Britain withdrew from India in 1947 the matter was still unresolved.[29] In Xinjiang, the situation was more chaotic, authority exercised by a succession of rapacious and brutal leaders.

After the Communists assumed power in China in 1949, Xinjiang and subsequently Tibet were re-absorbed into the newly established People's Republic of China (PRC). Initially the Tibetan government tried to negotiate a compromise, acknowledging Chinese sovereignty in return for assurances that the traditional socio-political system would be maintained. However, they were frustrated in their efforts. The Dalai Lama fled the country in 1959. The Tibet Autonomous Region (TAR), in Chinese called *Xizang*, was established in 1965 in western Tibet, on the territory that had formerly been ruled directly by the Dalai Lama and his government; there are also areas with Tibetan autonomous status in the neighbouring provinces of the PRC.

Soviet control of the southern region of western Turkestan was established (not very securely at first) in 1918. It was at this time that the Turkestan Autonomous Soviet Socialist Republic (ASSR) was created within, and subordinate to, the Russian Soviet Federative Socialist Republic (RSFSR). In the steppe zone Soviet power was not established until 1920, when the Kazakh (originally called Kirghiz) ASSR was created, which, like the Turkestan ASSR, was within and subordinate to the RSFSR. Likewise in 1920, People's Republics were proclaimed in the still nominally independent states of Bukhara and Khiva. In 1924 the Turkestan and Kazakh ASSRs, along with the newly annexed People's Soviet Republics of Bukhara and Khiva, were repartitioned along ethno-linguistic lines. As a result of this division (known as the 'National Delimitation of Central Asia'), five new administrative-territorial units were formed, namely the Uzbek and Turkmen Soviet Socialist Republics (SSRs), the Tajik ASSR (transformed into a full SSR in 1929), and the Kazakh and Kirghiz ASSRs (transformed into SSRs in 1936). Over the following years some relatively small changes were made to the internal borders between these republics, but otherwise they have remained unchanged. After the collapse of the Soviet Union at the end of 1991, they became the international boundaries between the independent Central Asia states.

In 1947 India and Pakistan acquired independence from British colonial rule. The status of Kashmir was left undecided, jurisdiction over it contested by the Indian and Pakistani governments. Afghanistan had never been fully under British control, although in some matters – notably freedom to pursue an independent foreign policy – it had been forced to accept certain limitations. It disencumbered itself of this constraint in 1919. Treaties of Friendship were concluded with Soviet Russia and Turkey in 1921, and shortly after with Iran and several European states. In 1934 Afghanistan joined the League of Nations. Despite pressure from its neighbours, Afghanistan mostly succeeded in adhering to its avowed principle of neutrality. In the 1970s, however, it was drawn into an increasingly close relationship with the Soviet Union. When Soviet troops invaded the sountry in 1979 to shore up the incumbent Marxist government, many in the West feared that this heralded the start of a new wave of Russian expansionism. In fact, the Soviet army was unable to defeat the Afghan opposition forces and withdrew from the country in 1989. Thereafter the country plunged into civil war and today is still being torn apart by warring factions.

By the late 1980s Soviet power in eastern Europe was rapidly disintegrating. The Mongolian People's Republic, too, began to distance itself from the Soviet Union. In 1989, Soviet troops were withdrawn from the MPR. In 1990 it became the first Asian socialist country to hold free, multi-party parliamentary elections. A hotly contested presidential election was held in 1993. A new constitution was adopted in 1992; the name of the country was officially changed from the 'Mongolian People's Republic' to 'Mongolia'. Relations with China have shown a marked improvement in recent years, though areas of sensitivity remain on both sides. Relations with the Russian Federation are still strong, especially as regards trade. In general, Mongolia has adopted an outwardlooking policy and seeks to broaden its economic and diplomatic ties, likewise educational, cultural, scientific and organisational links with Asian as well as with European countries.The ex-Soviet Central Asian states are at an even earlier stage of their development as independent countries than is Mongolia, but to date they have shown a similar commitment to following non-aligned policies, and a readiness to look east as well as west, north as well as south for potential partners. In former Soviet Central Asia, moves to create a Central Asian economic union were initiated shortly after the collapse of the Soviet Union, but have not as yet proceeded very far.

To sum up the political alignments at the end of the 20th century, most of eastern Central Asia (Xinjiang, Tibet and Inner Mongolia) forms part of the People's Republic of China; however, there are active separatist movements in Tibet, parts of Xinjiang and possibly, though less visibly, elsewhere. In western Central Asia, as also in Mongolia, there are states that have only recently acquired independence and are still in the process of coming to terms with the challenges, domestic and foreign, of this new development. In the north, the former Soviet administrative entities of Siberia still form part of the Russian Federation, but with a more sovereign status than under Soviet rule. In the south, Afghanistan remains in turmoil, and Kashmir continues to be disputed territory; the Himalayan mountain kingdoms are to a greater or lesser degree under Indian control.

Cultural Geography

There are many aspects of Eurasian culture that could be mapped, such as way of life; forms of social organisation; arts and crafts; iconography and design; music and dance; language groups; and material culture (including food, jewellery and dress). A study of each such feature would reveal a different set of linkages and cleavages, provide a different interpretation of the 'typical' markers of Central Asian culture. Such a survey, however, would require a whole book. In this section only two features will be considered, namely, religion and script, as an illustration of the complexity and fluidity of the contours of the cultural map of the region.

Natural religions

A profusion of faiths have flourished in Central Asia. Most of the main credal religions have been represented here, as well as various forms of 'natural religion' (i.e. shamanism and animistic cults). The Turkic and Mongol dwellers of the steppe and forest were adherents of this latter complex of belief systems.[30] Very little is known of the exact nature of the spiritual world of the earliest inhabitants of these regions, but there are marked resemblances in the burial customs and grave structures, and also in the animal-style decorative motifs that are found in a broad band that stretches from the Pontic steppe across southern Siberia and Kazakhstan to Mongolia, dating from the early 1st millenium BC.[31] The ancient Iranian nomads possibly worshipped pre-Zoroastrian divinities. For a much later period (from the 6th century AD onwards), the practice of shamanism and of animistic rites is well attested among Mongol and Turkic tribes; a widespread feature of their belief system was the cult of the sky divinity (*Tengrism*). In Tibet, the Bon religion, with its emphasis on animism and magic, was also a form of natural religion. The spread of credal religions gradually submerged – often eradicating by force – the natural religions of the steppe peoples. Yet in many cases these survived, usually in a semi-disguised form, sometimes renamed and re-interpreted so as to conform to the new religious norms.

Early credal religions

Credal religions were introduced into the southern tier of Central Asia from the surrounding sedentary states. The chief conduit was Iran, but India and China were also influential. The new faiths generally spread along the main trade routes, carried from one settlement to another by merchants and artisans, as well as by missionaries and refugees fleeing from religious persecution elsewhere. For much of the 1st millenium AD several religions co-existed in close proximity, as local rulers were, for the most part, eclectic in their religious affiliations and tolerant towards different belief systems. The period was also characterised by a strong tendency to religious syncretism. From about the 10th century AD onwards, however, two religions came to predominate, Buddhism and Islam; the geographic boundaries between them gradually crystallised and the modern religious map of the region took shape. The Tibetans and Mongolians adhered to Buddhism; the Iranian peoples and most of the Turkic peoples (at least nominally) adopted Islam.

The role of the Sogdians in the dissemination of religious beliefs deserves special mention. An Iranian people who originated in the Zerafshan valley (modern southern Uzbekistan and Tajikistan), they established communities to the west, but also (by the early 4th century AD) to the east of the Tien Shan. When transcontinental Eurasian trade was at its most flourishing, they were the pre-eminent entrepreneurs, controlling the east-west, north-south flow of goods in the key region between the Aral Sea and China. Their language was for centuries the lingua franca of a large part of the Silk Roads. They were also technocrats (specialists, for example, in the construction of irrigation channels, in horticulture, metalworking, glass-making and weaving) and artists of distinction. Culturally and economically, they were at the height of their powers in the 6th to 8th centuries AD, as is beautifully illustrated by the wall murals and other archaeological finds dating from this period from Samarkand and nearby settlements. They recorded and translated important canonical texts relating to several of the major religions.[32]

The earliest credal religion to make an appearance in Central Asia was Zoroastrianism. There is some disagreement amongst scholars as to the time and place of its origin. It probably arose in the 7th century BC, in eastern or north-eastern Iran, and thereafter underwent a long and complex development.[33] During the Achaemenid period Zoroastrianism spread throughout the Iranian world, but although it became the dominant faith, it did not at this time oust other cults. Under the Parthians, Zoroastrianism gained in importance and under the Sassanians it became the state religion. During this period, if not earlier, it spread north of the Amu Darya into the fertile valleys of Transoxiana and across the Syr Darya to the edge of the steppe (modern southern Kazakhstan); also further eastwards to the Tarim basin, eventually reaching China. In the 6th century AD Samarkand was an important Zoroastrian centre, and remained so until the mid-9th century, when Islam gained ascendancy.

Prior to the introduction of Islam, three other credal religions reached Central Asia through Iran: Christianity, Manichaeism and, on a far more limited scale, Judaism. They spread from west to east at approximately the same time and following almost identical geographic paths. They shared, too, in part, a common motivation, since all three were subjected to some degree of persecution in Iran by the Zoroastrian Sassanians; the Christian and Manichean emigrants, however, were also impelled by missonary zeal to propogate their religions.

Christianity, in the form of the Syriac church, was quite widely represented in Iran by the end of the 3rd century AD. Doctrinally, the Iranian communities favoured the Nestorian school of theology. Christian missionaries from Iran to Central Asia also espoused Nestorianism. They appear to have been active from the second half of the 4th century AD; bishoprics had been established in Merv (modern Mary) and Herat by the early 5th century. The religion spread across the Amu Darya into Transoxiana; a metropolitan see was created in Samarkand (the precise date is unknown, but not later than the early 8th century). Christian communities are also known to have existed in the Chu basin (modern Kyrgyzstan), and east of the Tien Shan. A concerted attempt to expand the Nestorian church took place under Patriarch Timothy (AD 780–823), who sent many missions to the east.

The religion seems mainly to have followed the northern branch of the Silk Road (see **Economic Geograpy** below, under *Trade: the Silk Roads*). Metropolitan sees were established in Kashgar (before the end of the 12th century) and Almaliq, and a bishopric in Hami; there was also a Christian colony in Dunhuang in the 8th to 10th centuries. Attempts to evangelise Tibet made no significant impression, but Nestorian Christians were certainly represented at the court of the Mongol khans in Karakorum by the 13th century. Moreover, a number of Christian emissaries from western European courts travelled across Central Asia at this time on diplomatic missions to the Mongol rulers, thus there was some contact with other forms of Christianity. Gradually, however, the Christian communties in Central Asia contracted, often as a result of local persecution. There was little Christian presence in the region beyond the 14th century.[34]

Manichaeism entered Central Asia at approximately the same period as Christianity, following much the same route. Mani, the founder of the faith, was born in Mesapotamia in AD 215/6. In the mid-3rd century, drawing on a variety of sources, particularly Mithraism, Zoroastrianism and Christianity, he formulated a new religion. Initially, Manichaeism enjoyed the support of the Sassanid ruler Shahpur I. His successors, however, favoured the Zoroastrian hieratic establishment and tried to eradicate Manichaeism; Mani himself was executed in AD 276/7. Nevertheless, the new faith continued to spread; it was established in Parthia before the end of the 3rd century and from there penetrated Transoxiana and the Tarim basin, mainly following, like Christianity, the northern branch of the Silk Road. It reached China before the end of the 7th century. It was at first welcomed at the Chinese court, but a few decades later suffered persecution.

However, non-Chinese subjects were allowed to continue practising the faith and it was through contacts with Sogdian Manichaeans in China that the Uighurs, a semi-sedentary Turkic people of Central Asia, were converted to this religion. In AD 763 it was proclaimed the official religion of the Uighur state. In choosing Manichaeism, the Uighurs also favoured Sogdian culture; this in turn led to the development of a more settled existence and the beginnings of an urban way of life. The Uighurs were overthrown and dispersed by the Yenisei Kyrgyz in AD 840 and this destroyed the state infrastructure of Manichaeism in the steppe. Nevertheless, the religion survived in the Uighur principalities in Gansu and the Turfan oasis until the 13th century. Elsewhere in Central Asia it disappeared somewhat earlier.

Judaism was also represented in Central Asia, though to a far lesser extent than the other major religions. Jewish merchants and artisans, probably from Mesapotamia, and most likely connected with the silk trade, settled in Merv in about the 4th century AD. There were Jews in Khwarazm (ancient Chorasmia, south of the Aral Sea) before the Arab invasion and by the 8th century in the Tarim basin. Under Muslim rule the Jews were treated as an autonomous religio-ethnic group, subject to the same laws and taxes as other non-Muslims. By the 12th century (if not earlier) there was a sizeable Jewish community in Samarkand. There were also Jews south of the Amu Darya (modern Afghanistan). In the 16th century, with the rise to power of a new dynasty in Transoxiana (the Uzbek Sheibanids), Bukhara became the focal point of Jewish life in the region. Treatment of the Jews in Bukhara varied from outright persecution to limited toleration. During the Soviet period they suffered from both religious and ethnic discrimination. Since the 1970s there has been massive emigration to Israel.[35] Communities of Bukharan Jews still exist in Central Asia, but now number no more than a few thousand.

Buddhism

Buddhism, one of the two great religions that was to make a lasting impact on Central Asia, first entered entered Central Asia through India. The historic founder of the religion, Sakyamuni, lived in northeast India at the end of the 6th century – early 5th century BC. Over the next two hundred years his followers codified his teachings into canonical texts and rituals. During the 3rd century BC, Buddhism was established in Bactria and Gandhara (modern Afghanistan and northwest Pakistan); at about the same time, Buddhist missionaries began proselytising along the southern margin of Central Asia, crossing the Pamirs into the Tarim basin, and thence on to China. Khotan became a major outpost of the Buddhist faith.[36] From there, a chain of stupas and monasteries was gradually established along the arteries of the Silk Road, extending across the Lop Nor desert to Dunhuang.

In the 2nd century AD, under the Kushan emperor Kanishka I, the spread of Buddhism in Central Asia received a fresh impetus. Although Buddhism was not the state religion (Iranian, Greek, Roman and Hindu deities were also venerated), it enjoyed the personal patronage of the ruler. It was during this period that there was a flowering of the distinctive Gandharan Buddhist school of sculpture, a synthesis of Greek and Indian aesthetic canons. After the fall of the Kushan empire Buddhism was persecuted by conquerors (Sassanians, later Hephthalites, finally Arabs) who esposed other faiths but the religion survived in Gandhara in such centres as Bamiyan and Hadda (modern Afghanistan) for several centuries.

It was during the Kushan period, too, that Buddhism began to penetrate Transoxiana, and from thence to spread westwards to the Murghab valley (modern Turkmenistan). Somewhat later, the faith was established in the southern steppe, possibly by missionaries from the Tarim basin; a significant Buddhist colony was located in the Chu valley (modern Kyrgyzstan), from the 6th to 10th centuries AD.

Further east, Tibet was also being drawn into the Buddhist orbit under Indian and Chinese influence; in the late 8th century AD, at a time when the country was relatively stable and united, Buddhism was proclaimed the state religion. The systematic translation of Buddhist scriptures was commenced; also, efforts were made to eradicate the Bon faith, the older, natural religion of Tibet. Thus, by the end of the 8th century AD, Buddhism was established throughout southern Central Asia, embracing, in modern terms, part of Afghanistan and north-west Pakistan; south-eastern Turkmenistan, parts of Uzbekistan, Tajikistan and Kyrgyzstan; and most of Xinjiang and Tibet.[37]

Buddhism subsequently both lost and gained ground in Central Asia. In Tibet, in the first half of the 9th century, the monarchy disintegrated and the state fragmented into local fiefdoms. The Buddhist clergy had become deeply unpopular and this provoked fierce opposition to the religion. Buddhism began to be persecuted and all but disappeared from Tibet. In the mid-11th century, however, there was a revival of the faith, under the guidance of Indian teachers. This was accompanied by a proliferation of monasteries and also of sects. A powerful and wealthy religious aristocracy emerged, paving the way to the formation of a theocratic state. This was further encouraged by the Mongols who, in the second half of the 13th century, held suzerainty over Tibet and appointed one of the abbots as the regional governor.

Buddhism gained a foothold in Mongolia at this time, but it was not until the mid-16th century that it was officially adopted as state religion, following the reformed school of Lamaism ('Yellow Hat' sect) that had been established in Tibet at the beginning of the 15th century. The third Dalai Lama is credited with the conversion of the Mongols (the title *Dalai*, Mongolian for 'Ocean', was bestowed on the Tibetan spiritual leader by the Mongol khan in 1568). In the religious sphere, the two countries have remained closely linked from this period onwards up to the present.[38] The neighbouring Turkic Siberian peoples who were under Qing rule (e.g. Tuvinians) were likewise converted to Lamaist Buddhism during the 18th century. Elsewhere in Central Asia, the influence of Buddhism waned; in Afghanistan and Turkestan (eastern and western) it was eventually superseded by Islam.

Islam

Islam, the second of the two defining religions of Central Asia, was brought to Central Asia by Arab armies. By the mid-7th century AD, within scarcely three decades of the *Hijra* (AD 622, the year of Muhammad's flight from Mecca to Medina, the formal commencement of the Muslim era) Muslim Arabs had conquered Khorasan and established a regional administrative centre at Merv. An expeditionary force made a first sortie across the Amu Darya in AD 667 (AH 45), but it was not until the early 8th century that the conquest of Transoxiana was undertaken in earnest. Once this had been accomplished, strenuous efforts were made to Islamicise the local population (Iranian and Turkic). Initially, this was undertaken by force, and accompanied by harsh persecution of other faiths, especially Zoroastrianism. Gradually, however, more moderate policies prevailed and by the 9th century the new religion had taken root in the urban centres of Transoxiana; by the early 11th century it was becoming established in the Tarim basin.

The necessary institutional infrastructure was created in terms both of human resources and of material facilities for communal worship and study. Scholars from the region travelled widely throughout the Muslim world; some made impressive contributions to Islamic philosophy and jurisprudence, also to mathematics, astronomy and other sciences, such as, for example, al-Bukhari, at-Tirmizi and al-Khwarezmi in the 9th century AD, and al-Farabi, al-Biruni and Ibn Sina (Avicenna) in the 10th century AD. Muslim culture and learning in the settled areas of southern Central Asia was resilient enough to survive the Mongol conquest in the 13th century. Indeed, the new rulers were themselves converted to Islam within a few generations and culturally this region remained within the Islamic orbit.

Islam took far longer to reach the peoples of the steppe. The members of the ruling aristocracy of the tribal confederations were the first to accept the faith; Kazakh khans from the 15th century onwards were keenly aware of the political significance of a link with the Islamic establishment of the south. The nomads of the north-east (modern Kyrgyzstan) were probably not completely Islamicised until the 18th century, and some of the outlying Kazakh tribes not until the 19th century, after their incorporation into the Tsarist empire.

The great majority of the Muslims of western as well as eastern Turkestan were, and still are, followers of orthodox Sunni Islam of the Hanafi school (like most of the rest of the Muslim world, but unlike Iran, where Shiism prevails). In the high mountains of the south-east, however, many of the Pamiri peoples (Shugnis, Wakhis etc.) were converted to the Ismaili sect of Shia Islam by missionaries from Afghanistan and India in the 11th century AD. Sufism, the mystical aspect of Islam, penetrated the oasis belt of southern Central Asia in the immediate aftermath of the Arab invasion. Important centres of Sufism developed in Merv, Bukhara and Khwarezm. The early adepts were followers of the Baghdad school of mystics. In the early 12th century, indigenous schools began to appear. The first major figure in the development of Central Asian mysticism was Yusuf Hamadani (AD 1048–1141), who spent most of his adult life in Merv. Other great Central Asian Sufis, founders of 'paths' (tariqa) that were in time to spread far beyond the confines of Central Asia, were Ahmad Yasavi (mid-12th century), Najm ad-Din al Kubra (AD 1145–1221) and Baha ad-Din an-Naqshbandi (AD 1318–89).

Religion in the 18th–20th centuries

The expansion of the Tsarist empire in Central Asia was accompanied by the introduction of Orthodox Christianity into some regions. The peoples of the Siberian forests and tundra (e.g. Yakuts, Khakass, Shors), originally shamanists, were evangelised by Russian missionaries from the 18th century onwards. Later, after the annexation of the steppe, an attempt was made to convert the Kazakhs to Christianity. This, however, met with little success and was soon abandoned. When the southern tier was conquered (modern Uzbekistan and Turkmenistan), there was virtually no attempt to spread Christianity among the indigenous population. Muslim institutions were allowed to continue functioning with little interference from the colonial administration.

Under Soviet rule, the situation changed radically. A determined attempt was made to eradicate all forms of religion. Places of worship were desecrated and closed, holy scriptures destroyed, clergy and religious teachers imprisoned or executed, and believers – or those merely suspected of being believers – were subjected to harassment and discrimination. The persecution was at its height during the 1930s, but even after the worst excesses had ceased, the legacy of fear, combined with a ceaseless barrage of atheistic propaganda, continued to erode religious faith. As a result, within a few decades, Islam in Soviet Central Asia had been largely reduced to a cultural identity.[39] In the MPR, Buddhism suffered a similar fate.[40] In the PRC, too, especially during the Cultural Revolution, religion was severely persecuted. Nevertheless, the process of secularisation did not proceed as far in Xinjiang as it did in the Soviet Central Asian republics; in Tibet, it met with even less success.[41] In both Xinjiang and Tibet, religion is still a powerful, mobilising force in society.

Since the disintegration of the Soviet bloc, there have been religious revivals in former Soviet Central Asia and in Mongolia. The main focus of attention has been Islam in the ex-Soviet states and Buddhism in Mongolia.[42] However, there has also been some revival of Tengrism and shamanism in the latter, and, in a small and informal way, of shamanism in parts of Kazakhstan. To some extent these movements are spontaneous manifestations of religious regeneration, but they are also motivated by a desire to re-establish links with the past, to recover a long repressed identity. There is, moreover, significant government support for the promotion of these 'national' religions, as the new administrations seek to replace Marxist dogma with a spiritual and ethical outlook that is rooted in the history and culture of their countries. However, attitudes towards religion remain ambivalent throughout the region. The current religious revival is welcomed, but it is also feared as a potential threat to the secular political systems that exist in these states. Such concerns are particularly strong in areas where Islam is the dominant religion. Consequently, in Uzbekistan and the other ex-Soviet states religious activities are closely monitored and subjected to a considerable degree of government control.

Scripts

There is a long history of the written word in Central Asia, characterised – as so much else in the region – by a multiplicity of forms. A single language has often been recorded in a number of different scripts; equally, a single script (appropriately modified, where necessary) has often been used for languages of different origins. The situation is further complicated by the fact that multilingualism was common, especially in the more densely inhabited regions of the south. Often, the choice of script has been dictated by religious and/or political affiliation rather than by linguistic considerations. Thus, the visual representation of a language acquired a symbolism that went far beyond the basic task of conveying particular sound values.

Plotting the spread of scripts in Central Asia amplifies the cultural map of the region and provides yet further illustrations of the constantly shifting spheres of influence. In other societies that have an established literary tradition the problem of selecting a writing system was resolved at some point in the distant past and is no longer a matter of debate. In Central Asia, however, it is still a live issue. The past century and a half has been a time of rupture and upheaval in the region and one of the ways in which this has been reflected has been in the changes of script. Today, as the newly independent states of Central Asia seek to redefine their identity, the choice of script for the national language is regarded as a question of fundamental importance precisely because the script itself is felt to be a proclamation of cultural and political orientation. The following survey of writing systems in Central Asia is of necessity brief. Nonetheless, it will perhaps suffice to show something of the complexity of the subject.

Early writing systems

The first introduction of writing into Central Asia probably occurred after the Achaemenid conquest. The official written language and script of the empire was Aramaic and it must surely have been used in the chancelleries of the eastern satrapies (i.e. on the territory of modern Afghanistan, Turkmenistan and southern Uzbekistan). However, no documentary evidence from this period has survived though some later Aramaic inscriptions have been found on the territory of Bactria and Gandhara. Also, several of the scripts that were subsequently used in Central Asia were derived from Aramaic.

A language and script that has been more fully preserved is Greek. Alexander the Great's Central Asian campaign was followed by a wave of Greek colonisation, particularly in Bactria. This resulted in a high degree of Hellenisation. The Greek language and script gained wide currency at this time. The surviving texts reveal a level of erudition and stylistic elegance that is in no way inferior to that of the Hellenic kingdoms of the Mediterranean. In the successor Greek states in Central Asia, especially in Graeco-Bactria, Greek continued to be used in all spheres of public life.[43]

The Kushan empire, which succeeded the Graeco-Bactrian state and encompassed a far greater territory, was eclectic in its attitude to language, art and religion. Three scripts were widely used. One was the Greek script, which, with minor modifications, was adopted for Bactrian, an Iranian language that had previously had no written form. Several Bactrian texts have survived, including an important inscription found near Termez, on the north side of the Amu Darya.[44] Brahmi, an Indian script, was used for Sanskrit and other Indian languages. Kharosthi, derived from Aramaic, was used for Gandhari Prakrit (also an Indian language). The use of Gandhari Prakrit, in the Kharosthi script, spread rapidly, carried eastwards along the Silk Road by Indian merchants and scribes, also by Buddhist monks. It was used for secular as well as religious purposes. By about 100 BC it penetrated northern Bactria and from thence, in the later Kushan period, was widely disseminated throughout the Tarim basin.[45]

Brahmi, too, was used at this same period for writing Gandhari Prakrit and like Kharosthi, also spread to the Tarim basin. Later, however, towards the end of the 2nd century AD, Brahmi superseded Kharosthi and became the dominant script in the Tarim basin. It was used for Buddhist Hybrid Sanskrit (which replaced Gandhari Prakrit as a literary language of Buddhism); also, slightly adapted, for Iranian languages such as Tocharian (vicinity of Kucha and Turfan), Saka (Tumshuk, near Kashgar) and Sogdian, likewise for Uighur, a Turkic language (Turfan).[46] The literature in these languages is mainly of a religious nature, including translated extracts from the Buddhist canon, commentaries and other edifying material.

Manichaean sacred texts, originally written in Aramaic and Middle Persian, were translated into several Central Asian languages, including Parthian, Bactrian, Saka (Tumshuqese), Tocharian, Sogdian and Old Turkic. They were transcribed in various scripts: Sogdian, Brahmi, Runic and also the Aramaic script used by Mani, which had acquired hieratic significance for his followers.[47]

The Sogdians used several scripts, depending on time, place and purpose. The one that is most specifically associated with the Sogdians, and is generally referred to as the 'Sogdian script', was derived from the Aramaic. It was widely used for secular as well as religious purposes for several centuries. The Sogdian script was subsequently adopted by the Uighurs when they converted to Manichaeism. Several centuries later, this script was transmitted to the Mongols by Uighur scribes in the service of Genghiz Khan; it was adopted as the national script and remained in use in Mongolia up to the 20th century. In the early 17th century, during the consolidation of Manchu power, it was likewise adopted as the state script of the Manchus and has continued to be used up to the present day.

The nomad Turks of the steppe also had their own form of writing. Probably derived from Aramaic via Sogdian, it is known as the Runic script because of its visual, though entirely coincedental, resemblance to the Scandinavian runes.[48] Very few examples of this script have survived. The majority take the form of short funerary inscriptions carved on stone slabs or rock faces. They date from the 5th to 8th centuries AD and have been found in the Chu-Ili basin and in Mongolia. The most extensive texts, located near the Orkhon river, date from the mid-8th century. They are formulated in a highly accomplished literary style, suggesting that they emerged out of a developed tradition of chronicle writing; however, no other examples of inscriptions of this quality and length have as yet been discovered. In Tibet, the development of writing was closely linked to the introduction of Buddhism. Initially, a form of Sanskrit was used; in the mid-7th century AD, however, an alphabetic writing system was created for Tibetan by Indian Buddhist monks and this remains in use.

The expansion of Islam was accompanied by the spread of the Arabic script, the script in which the Holy Quran was revealed. As first the peoples south of the Amu Darya (modern Afghanistan) were Islamicised, then of Transoxiana and the Tarim basin, and finally of the steppe region, so, too, the writing systems that had previously been used in these areas were abandoned in favour of the Arabic script.[49] Over the following centuries, especially in Transoxiana, a large corpus of literature, in a variety of genres, accumulated in the Arabic script. By the time western Turkestan was incorporated into the Tsarist empire an extensive indigenous literature existed both in Persian (Tajik), and in Chagatai (Old Uzbek), the main Turkic literary medium of Transoxiana. There was also some literature in the other Turkic languages of the region, including some poetry in Turkmen. However, there was very little written material in Kazakh and virtually nothing that could be termed Kyrgyz.

Scripts in the 20th century

In the 20th century, after Soviet power had been established, the languages of the main ethnic groups were given the status of 'national' languages and great efforts were made to develop their functional capacity. The Arabic script was reformed, but retained until the late 1920s. Thereafter it was abolished and the Latin script introduced in its place; a slightly different alphabet was devised for each of the languages, reflecting their particular phonetic features. Some of the Siberian Turkic languages (e.g. Yakut) had been written in the Cyrillic script from the 19th century onwards; these, too, were now given alphabets in the Latin script. In the period c.1924–30 a considerable quantity of educational, political, technical and literary material was published in all these languages. Nevertheless, in 1940 it was decreed that for all the Central Asian and Siberian languages the Latin script was to be abandoned forthwith in favour of the Cyrillic; in the MPR, the Mongol script was replaced by the Cyrillic the following year, in 1941. Various reasons have been suggested for this sudden move to the Cyrillic script, and for the extraordinary haste with which it was implemented, but the chief motivation was probably ideological, prompted by a desire for national consolidation in the face of the imminent threat of global war.

In the post-Second World War period there was an even sharper rate of increase in the volume and variety of publications in the Central Asian languages than there had been in the 1930s.[50] The Cyrillic-script alphabets appeared to be functioning satisfactorily. However, in the 1980s, particularly amongst Uzbek and Tajik intellectuals, demands began to be made for the re-introduction of the Arabic script, at least as a subject to be taught in schools. This was prompted not so much by a rejection of the Cyrillic script as by a desire to rediscover the cultural legacy of the past. In the MPR, too, there were similar calls for a revival of the Mongol script.

After the disintegration of the Soviet bloc in the early 1990s, the re-introduction of the Mongol script in the MPR began to be implemented. However, in the four predominantly Turkic Central Asian states (i.e. not including Tajikistan), the decision was taken to return not to the Arabic script but to the Latin, on the grounds that this script was more 'progressive' and signalled a commitment to political and economic reform; it would also bring these countries into line with Turkey, where the Latin script has been used since the end of the 1920s. Yet in Mongolia as well as in the ex-Soviet states, the change of script is proceeding slowly and not without some misgivings. Not only is there now full literacy in the Cyrillic script, but there is also a very large corpus of literature in that script in all these countries. There are concerns that a change of script will render at least a portion of the population illiterate; also, that much of the history and culture of the last half century will be lost to future generations.

In the PRC there were some attempts to change scripts, but these were short-lived.[51] Thus, at the end of the 20th century the distribution of scripts in the Central Asia is as follows: the Arabic script (different alphabets) is used in Afghanistan and by the Muslim peoples of Xinjiang; the Mongol script is still used by Mongols in the PRC and is being re-introduced into Mongolia; the Tibetan and Manchu scripts are used by the eponymous peoples; the Latin script is being re-introduced for the state languages in Kazakhstan, Kyrgyz-stan, Turkmenistan and Uzbekistan (slightly different alphabets for each), though it is only in the last two that any significant progress has as yet been made in this direction. For the present, the Cyrillic script continues to be used in these four states and in Mongolia; it is also still used in Tajikistan (though there are some calls for a move to the Arabic script), and for the Turkic languages of Siberia (e.g. Yakut).

Economic Geography

Only the two aspects of the economic geography of Central Asia most relevant to the present book will be considered here, namely, the nomad/sedentary symbiosis, which has been a central feature of the region throughout its history; and long-haul trade networks, which have at various periods provided the means of integrating the region into the global system of the day.

Nomadism and sedentary culture

The earliest inhabitants of Central Asia were settled agriculturalists, but for centuries it was the nomads who were the dominant element throughout most of the region. The domestication of the horse, and further north of the reindeer, made it possible to exploit the economic potential of the forest, steppe and semi-desert zones more efficiently through extensive animal husbandry than through farming. This led to the rise of a nomadic economy based on large-scale cattle-breeding throughout central Eurasia at the start of the 1st millenium BC. The general paucity of information about this period means that little is known for certain about the social organisation of these societies, but there does appear to have been a tight hierarchical structure, with several clan-tribal sub-divisions.

The horse proved to be invaluable not only for herding and transport, but also for warfare. The nomads acquired outstanding mastery of equestrian skills and used them to great effect in military operations. Mounted warriors from the steppe were to be the scourge of the sedentary world for many centuries. Their speed and mobility, coupled with strong internal organisation, enabled them to conquer immense areas in a very short space of time. Some of these nomad invaders subsequently became sedentarised and assimilated to the culture of the regions they had conquered. The territorial confines within which nomadic pastoralism was practised were not, however, much altered by these movements. In the steppe, forest and semi-desert this way of life was maintained, with little change, up to the modern period.

Sedentary culture developed around the outer periphery, primarily in the south. In the oasis-river belt there were already numerous settled communities by the 10th-8th centuries BC. Archaeological finds show that arable farming, metallurgy, crafts (e.g. weaving and the making of decorated pottery) and domestic architecture were well developed, especially in Transoxiana and Bactria. Urbanisation progressed as the influence of neighbouring empires increased, linking the region to highly sophisticated sedentary civilisations. Bactria, under Greek rule, became known as 'the land of a thousand cities'. Greek styles and architectural orders were used here, in combination with Achaemenid and neo-Babylonian forms, and indigenous construction technologies. Cities like Ay Khanum (modern Afghanistan) were provided with such typical Greek public buildings as gymnasia, theatres, pantheons and sanctuaries. Further north, in Transoxiana and the Tarim basin, Iranian (Sogdian and Saka), Indian and Chinese influences prevailed. The cities here were cosmopolitan trading entrepots, as well as multi-ethnic centres of culture and learning. During the Islamic period (9th century onwards), particularly in Transoxiana, there was a great flowering of science and the arts. Muslim architecture and principles of town planning, characterised by the distinctive groupings of mosques, maddrassahs, caravansarais and other socio-religious establishments, were introduced, providing a visual affirmation of the integration of this part of Central Asia into the world of Islam.

In Mongolia and Tibet, nomadic pastoralism was the predominant way of life, with farming and other sedentary or semi-sedentary activities playing a relatively small role. Nevertheless, there was some development of settled communities in these areas, too. This process was closely linked to the rise of Lamaist Buddhism and the foundation of great monasteries. These institutions were not only religious complexes, but substantial economic enterprises, serviced by a sizeable permanent work force. They fostered arts and crafts, especially related to the production of manuscripts and printed books, religious paintings and embroideries, and other cult objects. Trade also concentrated around these establishments, as they provided the few fixed points for an otherwise peripatetic society. Given the size of the monastic communities, food production was also a major undertaking, requiring a variety of agricultural and processing skills. Later, as townships developed, they were invariably centred around a monastery.

There had always been a certain amount of interaction between the settled peoples and the nomads, primarily through trade, but sometimes also through political and military alliances. There had also been some gradual encroachment on to nomad lands, as centres of settled activity increased and the area of land under cultivation increased. However, the overall demographic and economic balance between the two groups did not alter either rapidly or markedly. This changed in the modern period, when large portions of Central Asia were incorporated into neighbouring states (see section on **Political Geography** above). Great numbers of immigrants from non-nomadic societies began to settle in the region. Also, economic and administrative policies were implemented that in many cases were inimical to nomadism. The areas that suffered most from these developments were Inner Mongolia under Qing rule, and later Kazakhstan under Tsarist rule.[52] Nomadic pastoralism continued to be practised, but under increasingly difficult and restrictive conditions. In the early 1930s, in Soviet Central Asia collectivisation and mass sedentarisation of nomads was enforced. Similar policies were later enforced in the MPR and PRC, though in most places not as ruthlessly as in the Soviet republics. Consequently, in the MPR and PRC some degree of traditional nomadism was able to survive.

In the latter half of the 20th century the rate of urbanisation increased sharply throughout Central Asia, particularly in the (ex-) Soviet sector. There was some development of light industries as well as selected sectors of heavy industry (see **Geography of Fragile Environments** below). Employment opportunities diversified. These factors, along with the spread of modern social services, especially in health care and education, and the improvement in transport and communications systems, have metamorphosed Central Asian societies, affecting even those communities in which nomadic pastoralism is still practised. Today, a predominantly traditional form of nomadism, albeit with some elements of modernisation, is still found in Mongolia, western Xinjiang (Tien Shan mountains and foothills) and in western Tibet (TAR/Xizang). Elsewhere (e.g. in former Soviet Central Asia), even though animal herders continue to practise transhumance, the socio-economic organisation of livestock rearing has been transformed and consequently, the seasonal migration is now subject to considerable bureaucratic control. The culture – beliefs, practices, etiquette, crafts – that was once an inherent feature of the nomad world has almost entirely vanished in these areas. At the same time, a settled way of life has become increasingly widespread and it is this rather than nomadism that, in the late 20th century, is characteristic of most of Central Asia (at least in terms of population, if not in actual size of territory). Nevertheless, recently sedentarised peoples (e.g. the Kazakhs) still retain a folk memory of their nomad heritage and this continues to shape their sense of identity. They perceive themselves to be different in outlook and character from neighbours who have a long established sedentary culture (e.g. the Uzbeks), even though the two groups may now lead identical lives. Thus, whereas in the past the nomad/sedentary divide could be mapped in geographic terms, today it is as much a notional as a physical division.[53]

Trade: the Silk Roads

From earliest times, long-distance trade has been an important feature of economic activity in Central Asia. It was the fabric of transcontinental linkages, enabling contact to be made between nomads of the steppe and the settled communities of the oases, between the Eurasian interior and the distant metropolitan centres in China, India, the Middle East, and the Mediterranean littoral. The entire complex of ancient trade routes is generally referred to as the 'Silk Roads', though the term itself dates from the 19th century. A number of different itineraries were used, depending on environmental conditions and the contemporary political situation. The commodities that were traded varied, but in general, the nomads supplied raw materials such as skins, furs, horses and precious minerals, in return for manufactured goods, luxury wares and some agricultural produce (e.g. tea, sugar) from the sedentary peoples located on the periphery of the steppe as well as further afield.[54]

It is impossible to put a date to the commencement of long-distance trade along the Silk Roads. The earliest contacts seem to have developed in the south-west, possibly in the 3rd millenium BC, stretching from the Mediterrannean and Arabian Seas across the Iranian plateau to northern India and Transoxiana. Similar networks developed to the east of the Tien Shan, leading from the Tarim basin to China. The most valuable commodity that was traded along this route at this early period was Central Asian jade, prized for its beauty as much as for its usefulness for tool-making. By the early 2nd millenium BC jade was beginning to appear to the west of the Tien Shan, as far away as Mesapotamia, which would seem to indicate that the eastern and western routes were linked. The domestication of the camel, which occurred at about this period, proved to be as important for the development of long-distance trade as did the domestication of the horse for the development of nomadism. The camel's capacity for enduring long periods with little water or fodder, and its adaptability to different terrains and extremes of climate, made it an invaluable asset for these long journeys across barren uplands, arid deserts and high mountains. The horse, however, also played an important role in facilitating trade, providing transport for the human escort and traction for wagons and carts laden with goods and chattels.

There were several routes across Central Asia. The main way from China was through the Gansu corridor to Dunhuang. Thereafter, the road bifurcated round the oval-shaped Taklamakan desert, one branch passing through the oases of the northern rim such as Loulan, Kucha and Aksu, the other along the southern rim through oases such as Khotan and Yarkand. They joined up at Kashghar, then divided again to cross the Pamirs, one branch leading north into the Ferghana valley through Kokand and Khojand to Samarkand, two others following a more southerly direction before converging again at Balkh. From there, some routes led due west to Bukhara and beyond, others south across the Hindu Kush. A more northerly way led from the Great Wall to the oasis of Hami, then across the steppe to Talas and Tashkent. An offshoot of this route led through the Dzungarian gap across the steppe north of Lake Balkhash. Many smaller routes formed connecting networks between these main arteries, including some that extended into Tibet, Ladakh and Kashmir.[55]

The southern tier of Central Asia frequently formed part of the outer perimeter of larger realms. This hugely expanded the scope for contacts with more advanced economies, providing access to highly developed transport and communication systems and to organised trade and commodity production networks; the exchange of design and advanced technology was also greatly facilitated. These links had the most direct impact on Transoxiana, but they also penetrated far into the steppe. During the Achaemenid period, for example, objects from the Near East were transported as far north as the Altai mountains, as is evidenced by finds in the burial mounds of Pazyryk and Tuva. This was also the period when Central Asia began to be exposed to influences from the Hellenic world on the western edge of the Iranian empire. Greek art and economic power was disseminated to the east via the roads and provincial capitals of the Achaemenid state long before the campaigns of Alexander the Great. A money-based economy was already well established in the Greek city states by this time and Greek coinage began to penetrate southern Central Asia. Later, after the establishment of Greek kingdoms in Central Asia, the impact of Hellenic culture on the region increased, and this in turn helped to strengthen links with the Mediterranean economic system.

Political developments on the periphery of Central Asia as well as in the interior had a crucial influence on commerce. The emergence of a new nomad power in the steppe region was generally accompanied by the disruption of long distance trade as routes became unsafe and caravan trains were liable to be attacked. Once the new group had gained control of a critical mass of territory, however, stability was restored and trade recommenced. The sedentary empires around the rim sometimes tried to counteract this dislocation, extending their control over the main routes beyond the actual bounds of their lands by establishing outlying military bases and by building alliances with local rulers; the Chinese frequently resorted to such tactics in the Tarim basin. At times the frontiers of these sedentary empires were so extended that they became contiguous and consequently, between them they were able to regulate successive legs of the main branches of the transcontinental trade routes.

The golden age of the Silk Roads is often considered to be the period of the Great Kushans (mid-1st century AD to 2nd quarter of 3rd century AD). At this time Bactria, the pivotal central section of the east-west routes formed part of the Kushan empire; from here, branches of the Silk Roads spread eastwards across Chinese-controlled territory in the east, westwards across Parthia, and beyond that, to Rome – a total distance of some 10,000km. There were also connecting routes through Kushan lands that led south through the Indus valley to ports on the west coast of India. The most famous commodity of transcontinental trade was Chinese silk. However, many other goods were transported along these routes. Archaeological excavations in southern Central Asia have revealed something of the international character of trade in this period, yielding examples of coins and wares from all over the known world, including glass from the Middle East, ceramics from the Mediterranean, mirrors and other luxury goods from China.

Transcontinental trade across Central Asia continued for many centuries. It varied in intensity, in market orientation, and to some extent in routes. It was not until the 15th century, however, that it finally began to lose its significance. This was partly because the political fragmentation that followed the disintegration of the Mongol empire led to instability in the interior of the region and this increased the hazards of long distance trade. A more important reason, however, was the rise of maritime trade. Sea routes were economically more viable than the tedious, dangerous, terrestrial alternatives. Shorter haul trade with the surrounding countries continued to flourish, but the long-distance extensions that had once linked Central Asia to the coastal regions of Asia and Europe were gradually abandoned.

New networks

In the 17th century Russia began to emerge as an important trading partner for the western steppe, and eventually for the whole of western Turkestan. In the 19th century there was some competition from British India, but by the early 20th century the economy of the region had been fully integrated into that of the Tsarist empire. After the establishment of Soviet power, the links between western Turkestan (i.e. Kazakhstan, Kyrgyzstan, Uzbekistan Tajikistan and Turkmenistan) and the adjacent territories (Iran, Afghanistan, the Indian sub-continent, Xinjiang) were almost completely severed. Transport, communications and trade relations were developed in such a way as to bind the region comprehensively and exclusively into the Soviet Union. The MPR, as a Soviet satellite state, was also integrated into this autarkic economic complex. Somewhat later, after the Communists came to power in China, eastern Turkestan and the other parts of Central Asia that were under Chinese rule underwent a similar process of integration into the PRC, and a corresponding isolation from neighbouring countries.

In the past few years, however, since the collapse of the Soviet Union, old cross-border ties are being revived. The former Soviet

Central Asian states are re-establishing regional links, as well as developing new economic relations with more distant countries (e.g. in western Europe, the Middle East, north and south America, Australia and New Zealand). Mongolia, too, has made considerable progress in diversifying its foreign economic relations. There is great optimism in the region regarding the prospects of reviving the ancient Silk Roads. Some of the infrastructure is already in place (e.g. road and rail links between Kazakhstan and Xinjiang, and between Turkmenistan and Iran). Thus, at the end of the 20th century, after a hiatus of some five hundred years, it is again becoming possible to travel and to transport goods overland from the east coast of China to the Persian Gulf, and indeed, to the Indian sub-continent and to western Europe.[56]

Geography of Fragile Environments

The ecosystems of Central Asia, whether in the steppe, the deserts or the high mountains, are exceptionally fragile. In the past, the local inhabitants, nomadic and sedentary, treated the land with respect, husbanding it carefully to protect it from the damaging consequences of overuse. In the latter part of the 20th century, however, much of Central Asia has been intensively developed and as a result, its carrying capacity has been strained to the point of imminent collapse. The blame for this must in large measure be placed on the central governments that imposed unsuitable and unsustainable development programmes on these vulnerable regions. There are, though, some more general problems. One is the very real dilemma of how to reconcile the desire to develop a modern economy, with its expectations of continuing growth, with the natural constraints that are inherent in a very delicately balanced ecosystem. Another is that it takes time for the full effects of a given course of action to emerge: with hindsight it is not difficult to see why particular policies were disastrous, but when they were initiated the outcome was not always so easy to predict.

There are four main factors that have exacerbated the region's environmental problems: the rapid and massive intensification of agricultural output; the introduction of harmful and inappropriate industrial technologies; the steep demographic growth; the loss of traditional skills and knowledge. These developments were particularly marked in those parts of Central Asia in which socialist systems were established. It should not be forgotten, however, that even in regions that have not had socialist governments as, for example, the Himalayan states, similar problems have emerged as a result of over-hasty and inappropriate forms of modernisation. These issues are examined in other chapters in this book. This section will give an overview of the common problems.

Policies to intensify agricultural output were first introduced in Soviet Central Asia and the MRP in the 1930s, hence earlier than elsewhere in the region; they were also practised more comprehensively and more consistently for a much longer period and the level of environmental damage was correspondingly higher than in the surrounding areas. There were three main thrusts to Soviet agriculture in Central Asia: the expansion of the cotton crop (particularly in Uzbekistan and Turkmenistan); the introduction of grain cultivation into the 'Virgin Lands' (i.e. the hitherto unploughed steppe) of central Kazakhstan and Buryatia; and the increase in head of livestock, especially of sheep for wool production (Kyrgyzstan and Tajikistan). All triggered disastrous chain reactions.[57]

Efforts to raise cotton production resulted in a vicious circle of agricultural malpractices. These included the abandoning of crop rotation in favour of a monoculture; also, the over-use of heavy machinery, which compacted the earth and eliminated all the living organisms, turning it into a dense, inert mass. Practices such as these reduced the fertility of the soil and led to increased dependency on chemical fertilisers. There was, too, a vast increase in the use of water for irrigation. This resulted in the depletion of the two main rivers, the Amu Darya and the Syr Darya. The flow of water to the Aral Sea was consequently greatly diminished and the Sea began to dry up. The surface area of the Sea has been reduced by almost a half over the past three decades. This has not only resulted in a chronic water deficit, the loss of the bio-resources of the Sea, and a change in the regional micro-climate, but the exposed sea-bed itself constitutes a major hazard: sand as well as the toxic residue from the chemicals that were used to boost the cotton harvest are scooped up by vicious dust storms and deposited over the surrounding countryside, causing rampant desertification.

In Kazakhstan and Buryatia, the cultivation of cereals met with some initial success, but later severe soil erosion set in. This was caused by a combination of poor land management and unfavourable environmental conditions, such as weak soil structures, high winds and extreme ranges of temperature. In Kyrgyzstan and Tajikistan, the increase in herd and flock sizes likewise led to severe ecological damage;[58] there has been deforestation, as well as massive soil erosion and pasture degradation, the result of indiscriminate over grazing and over-trampling of the thin soil cover. Similar problems are to be found throughout the Central Asian interior. In Xinjiang, for example, there are acute examples of pasture degradation through overgrazing; this in turn has encouraged infestations of rodents and insects, which have further exacerbated the damage to the grasslands. In Inner Mongolia, campaigns to open the steppe to grain cultivation were implemented at approximately the same time, and with similar consequences, to the 'Virgin Lands' project in Kazakhstan.[59] In parts of Tibet, too, poor land management has caused degradation of the grasslands.

Industrialisation in Central Asia has to date been fairly limited. The main branches of heavy industry are linked to the extraction and processing of minerals. The technology used in these enterprises is invariably outmoded, inefficient and environmentally harmful. Consequently, in the immediate vicinity of industrial plants the levels of air, soil and water pollution are very high. However, a far greater threat is posed by the research and development activities carried out on the military facilities that are located in the region. These include the former Soviet nuclear test site at Semipalatinsk (Kazakhstan) and the Chinese nuclear test site at Lop Nor (Xinjiang). The space centre at Baikonur (Kazakhstan) is also regarded by some as a source of environmental damage.

The problem of excessive population densities is localised rather than spread across the entire region. In most cases, the traditional geography of settlement has been maintained: today, as in the past, the most heavily populated areas are those with good reserves of productive land and water; elsewhere, the population is very thinly spread. What has changed in recent times is that density levels in these areas have increased very considerably. In the MPR, for example, the population rose from some 655,500 in 1924 to 2.5 million by the early 1990s. In Soviet Central Asia, the population almost quadrupled in the period 1926–89, rising from 14.6 million to just under 50 million. Sharp rises in population have also been experienced in Inner Mongolia, Xinjiang and Tibet.

This rapid rate of increase has been largely the result of continuing high birth rates, along with falling infant mortality and longer life expectancy (owing to better health care, improved sanitation and hygiene etc). In Soviet Central Asia, however, there was an additional factor, namely, the heavy influx of immigrants from other Soviet republics. Inner Mongolia, likewise, has undergone a prolonged and massive influx of immigrants from central China, with the result that this region has long been culturally more Chinese than Mongol. Xinjiang and parts of Tibet[60] have also experienced major waves of immigration, especially in recent years. The southern margin of Central Asia (e.g. the Himalayan states) have not been exposed to such a high volume of immigration. However, here, too, there are high birth rates and longer life expectancy than in the past, hence the trend is towards rapid population growth.

The loss of traditional expertise is the fourth factor that has contributed to environmental degradation in Central Asia. It has been caused in great measure by the accelerated pace of modernisation. In the socialist countries, development policies were planned, funded and implemented by central governments. There was little consideration for local conditions or concerns. Large-scale projects took the place of the family enterprise: the skills and crafts of the individual had no place in this new world. It created an irreparable rupture with traditional society. Within a decade or two, an entire way of life – a way of relating to the environment – was lost. In some cases such schemes were imposed with great brutality. In Kazakhstan, the enforced sedentarisation and collectivisation of the nomad population in the early 1930s resulted in tremendous loss of human and animal life.[61]

Collectivisation policies were also introduced in the MPR at this time; however, resistance from the nomads was so fierce, and the consequences, in terms of productivity, so disastrous, that the programme had to be abandoned. A modified form of collectivisation was re-introduced in the 1950s and this met with more success; some forms of traditional land use and animal husbandry were preserved, though here, too, links with the culture and skills of the past were weakened. Collectivisation was likewise pursued in the western regions of the PRC after the establishment of Communist rule, frequently with painful results.

The Himalayan states in the south fared better, in that they were not subjected to a regime of totalitarian coercion. Nevertheless, they also experienced sudden, though very uneven, modernisation in many spheres of life. This process was driven from above, by their respective governments, or from outside, by international agencies, rather than growing organically from within. The result was that, as under the socialist systems, the experience of the past was devalued. The younger generation came to regard traditional knowledge as primitive, an obstacle to progress. Thus, here, too, there was a rupture with the past. Likewise, environmental damage was inflicted carelessly and unwittingly, because the instinctive understanding of the fragile ecological balance that had been nurtured by previous generations had been lost.

Environmental Awareness and Regional Identity

As this chapter has tried to show, the concept of 'Central Asia' is protean, its definition dependent on time and subject. It has no clear geographic limits. It encapsulates a huge range of social, cultural and historical experiences. Features such as religion, political organisation or economic structure may for a time have served to 'mark off' one area from another, but these borders have dissolved as systems changed and new boundaries came into being. Yet there are some common threads. Perhaps the strongest and most constant of these, shared by all who have ever inhabited this landlocked core of Eurasia, is the experience of having to come to terms with a challenging environment. Conditions in the region that lies between the Caspian and the Gobi, the steppe and the Himalayas, are in most places rugged in the extreme, not readily conducive to human settlement. Yet the ecological systems are singularly vulnerable, easily damaged by human activity. Here, merely in order to survive, the local population had to learn to treat the natural world with respect and circumspection. This imposed a certain discipline and rhythm on their way of life and even, some would claim, inspired a sense of spirituality.

Today, throughout all of this region, there is a growing realisation that traditional ways of managing the environment must be revived. This does not mean a rejection of the benefits of modern society and the very real improvements in living standards that have taken place in the 20th century. Rather, a synthesis is being sought that will combine the best practices of traditional society with those of today's world. At government level, and also through locally-based non-governmental organisations, these issues are being raised and projects are being undertaken that address such concerns.

In trying to cope with the legacy of the inappropriate development policies of recent decades, the peoples of the region are discovering a sense of solidarity, united by a determination to relearn the secret of working with, rather than against their complex environment. They are becoming aware that sensitivity to the natural world is a unifying factor amongst them, and moreover, a marker of identity that distinguishes them from 'outsiders' – the products of gentler, more moderate climes, who tend to be less acutely attuned to the fragility of the environment. Thus, the quest for sustainable modes of development is today very much a regional cause and in this sense may be considered to be a defining feature of 'Central Asia'.

Notes

1 Guy Imart aptly comments that 'Much like the Loch Ness monster, the dispute over the limits of "Inner Asia" surfaces periodically . . .' (G. Imart, *Limits of Inner Asia*, Bloomington, Indiana University Publications, 1987, p. 1). The same could be said of attempts to set limits to 'Central Asia'. Gavin Hambly concludes that 'as a geographical expression the term 'Central Asia' tends to elude precise definition' (G. Hambly *et al*, *Central Asia*, London, Weidenfeld and Nicolson, 1969, p.xi); André Gunder Frank describes it 'as a sort of black hole in the middle of the world' (A. G. Frank, *The Centrality of Central Asia*, Amsterdam,VU University Press, 1992, p.1). A glance at the contents of any dozen or so books with 'Central Asia' in their title will reveal almost as many different concepts of the region.
2 Agreed at a summit meeting of the Presidents of Kazakhstan, Kyrgyzstan, Tajikistan, Turkmenistan and Uzbekistan, held in Tashkent, January 1993.
3 V. V. Barthold, 'Turkistan', *Encyclopaedia of Islam*, eds M. Th. Houtsma *et al*, Leiden, E. J. Brill, 1987 (reprint of 1st edition), vol. 8, p. 895.
4 From time to time it was under Uzbek rule and still has a large Turkic population.
5 Yet it was retained as the designation of a Soviet military district (centred on Tashkent); it is also the name of a town in southern Kazakhstan.
6 See, for example, the writings of Baymirza Hayit, one of the most active Uzbek émigré scholars of the post-war period (e.g. *Turkestan im Herzen Eurasiens*, Studien Verlag, Cologne, 1980).
7 The *Oxford English Dictionary* (compact edition), Oxford, Clarendon Press, 1971, vol. 2, p. 3240 (102), gives a citation from Chaucer, dating from c. 1369.
8 See Yuri Bregel, *Notes on the Study of Central Asia*, Papers on Inner Asia no. 28: Indiana University Publications, Bloomington, 1996, p. 1; also, compare the limits of 'Inner Asia' as defined in C. E. Black *et al*, *The Modernization of Inner Asia*, M. E. Sharpe Inc., Armonck NY, 1991, p. 3 (i.e. Iran, Afghanistan, Mongolia, Kazakhstan, Kyrgyzstan, Tajikistan, Turkmenistan, Uzbekistan, Tibet and Xinjiang), with the overlapping definition of 'Central Asia' given in *History of Civilizations of Central Asia*, Unesco, Paris (hereafter *Unesco History*), 1992, vol. 1, pp. 477–80 (i.e. Afghanistan, western China, northern India, north-eastern Iran, Mongolia, Pakistan and the former Soviet Central Asian republics).
9 Herodotus, *Histories*, book IV, 42–45 (English translation A. D. Godley, W. Heinemann, London, 1928) noted: 'I wonder, then, at those who have mapped out and divided the world into Libya, Asia, and Europe . . . Nor can I guess for what reason the earth, which is one, has three names, all of

women . . . nor can I learn the names of those who divided the world, or whence they got the names which they gave'.

10 T. Barfield, *The Perilous Frontier: Nomadic Empires and China*, Basil Blackwell, Oxford, 1989, p. 2.

11 A good summary of the main arguments is provided in *Unesco History*, vol. 1, pp. 477–80.

12 M. Yapp, 'Tradition and Change in Central Asia', in *Political and Economic Trends in Central Asia,* ed. S. Akiner, British Academic Press, London, 1994, pp. 1–10, discusses terminological usage in the political sphere at this period.

13 D. Sinor, in *The Cambridge History of Early Inner Asia*, ed. D. Sinor, Cambridge University Press, Cambridge, 1990 (hereafter *Cambridge History*), pp. 1–18, argues persuasively for such an interpretation.

14 Approximately between longitude 15°E and 130°E; at the western and eastern margins approximately between latitude 40°N and 50°N, in the centre between latitude 25°N and 60°N.

15 For a more detailed description of the physical geography of the region, illustrated by geomorphic and geobotanical maps, see *Unesco History*, vol. 1, pp. 29–44; also *Cambridge History*, pp. 19–40.

16 There are numerous scholarly studies on particular topics, or particular geographic areas, but very few works indeed that present an overview of the larger 'Central Asian' region. This is not surprising, given the daunting array of languages (Chinese, Latin, Greek, Arabic, Armenian, Persian and other Iranian languages, several Turkic languages, and a clutch of modern European languages), as well as disciplinary skills (philology, archaeology and numismatics, to name but three) that are needed. Two works that did attempt to do this, and made a promising start, are the *Cambridge History* and the *Unesco History*. Unfortunately, neither has progressed beyond the early medieval period. Nevertheless, they are both extremely useful works, not least for their excellent bibliographies.

17 Herodotus is the most important early source for the history and customs of the 'nomad Scythians' (see esp. book IV, 1–31). The Persians used 'Saka' as a collective ethnonym for the Scythian tribes, while the Greeks used 'Scythian' to denote the Iranian nomads of southern Russia and the Pontic steppe. However, it should be noted that there is little scholarly agreement on the question of the geographic distribution of the Saka tribes (*Unesco History*, vol. 2, pp. 24–33). See also John R. Gardiner-Garden, *Ktesias on Early Central Asian History and Ethnography,* Papers on Inner Asia no. 6, Indiana University Publications, Bloomington, 1987, and *id., Greek Conceptions on Inner Asian Geography and Ethnography from Ephoros to Eratosthenes*, Papers on Inner Asia no. 9, Indiana University Publications, Bloomington, 1987.

18 Complementary studies of the Hsiung-nu are provided by *Cambridge History*, pp. 120–50, and *Unesco History*, vol. 2, pp. 151–69. See also *Cambridge History*, pp. 177–78.

19 *Unesco History*, vol. 2, pp. 247–54, and *Cambridge History*, pp. 161–71, summarise the main points of dispute; *Unesco History*, vol. 2, chapters 11–15, and 17, provide studies of the culture, social and economic systems

of the Kushans, drawing together the research of an international group of scholars.

20 There is general (though not unanimous) agreement that the Hephthalite language was Iranian (*Cambridge History*, p. 300). Quite a wide range of sources mention the Hephthalites and provide information on such topics as their script, religion and physical appearance; however, much of this is fragmentary and contradictory, hence it is not possible to construct a clear picture of their civilisation.

21 The classic study of this campaign is H. A. R. Gibb, *The Arab Conquests in Central Asia*, Royal Asiatic Society, London, 1923.

22 See further H. H. Howorth, *History of the Mongols*, 4 vols, Longmans, London, 1876; J. J. Saunders, *The History of the Mongol Conquests*, Routledge & Kegan Paul, London, 1971; B. Spuler, *History of the Mongols. Based on eastern and western accounts of the 13th and 14th centuries* (translated from the German by H. and S. Drummond), Routledge & Kegan Paul, London, 1972.

23 The term does not appear to have been used by the Mongols. The fourteenth-century traveller Ibn Battuta speaks of the Tatar-Mongol khan's 'golden tent', but the first known use of the 'Golden Horde' to denote a state (or people) dates from 1564, when it was used in a Russian chronicle, *Istoriya o Kazanskom Tsarstve* (in *Polnoye sobraniye russkikh letopisey*, XIX, St. Petersburg, 1903).

24 An interesting analysis of Tamerlane's political heritage is to be found in Beatrice Forbes Manz, *The Rise and Rule of Tamerlane*, Cambridge University Press, Cambridge, 1989.

25 The Kalmyks were deported from the Kalmyk ASSR to Central Asia by Stalin in 1943 for alleged treason; they were 'rehabilitated' and allowed to return home in 1957–58. Quite separately, a small group of Kalmyks settled in the vicinity of Lake Issyk Kul (modern Kyrgyzstan) in the 19th century; they adopted Islam and came to be known as 'Sart Kalmyks'. The main body of Kalmyks remained Lamaist Buddhists.

26 There are still small communities of Greeks and Arabs in modern Uzbekistan, but these are descendants of later immigrants. The Arab immigration almost certainly dates from the time of Tamerlane, while the Greeks (some deportees from the Black Sea, some political refugees from Greece) arrived during the Soviet period.

27 This is discussed further in S. Akiner, *The Formation of Kazakh Identity*, Royal Institute of International Affairs, London, 1995, pp. 22–28.

28 Halford Mackinder was alert to the strategic implications of this: 'Is not the pivot region of the world's politics that vast area of Euro-Asia which is inaccessible to ships, but in antiquity lay open to the horse-riding nomads, and is today about to be covered with a network of railways?' ('The Geographic Pivot of History', *Geographic Journal*, vol. 23 (1904), pp. 241–64, reprinted in *Democratic Ideals and Reality*, ed. A. J. Pearce, Norton & Co., New York, 1962, pp. 421–37).

29 Premen Addy, 'British and Indian Strategic Perceptions of Tibet', in *Resistance and Reform in Tibet*, eds R. Barnett and S. Akiner, Hurst & Co., London, 1994 (hereafter *Tibet*), pp. 15–50, gives a perceptive analysis of this issue.

30 For a description of shamanistic beliefs and practices in Mongolia (including a good bibliography), see M-D. Even, 'The Shamanism of the Mongols', in *Mongolia Today*, ed. S. Akiner, Kegan Paul International, London, 1991 (hereafter *Mongolia*), pp. 183–205; see also A. Molnar, *Weather Magic in Inner Asia*, Indiana University Press, Bloomington, 1994, for an excellent and wide-ranging study of Mongol and Turkic beliefs and rituals relating to the manipulation and control of the weather.

31 E. D. Phillips, *The Royal Hordes: Nomad Peoples of the Steppe*, Thames and Hudson, London, 1965, is a good starting point for those not familiar with this subject. It is well illustrated and includes several maps and a clear chronological table.

32 For further information on the Sogdians see R. E. Emmerick, 'Iranian settlement to the east of the Pamirs', in *Cambridge History of Iran* (hereafter *Cambridge Iran*), vol. 3/1, ed. E. Yarshater, Cambridge University Press, Cambridge, 1983, pp. 263–75; see also references to the Sogdians in other chapters.

33 Mary Boyce, *A History of Zoroastrianism*, vol. 1, Brill, Leiden, 1975, suggests a much earlier date (1400–1000 BC) and a location further to the north-east (p. 160).

34 An admirably concise survey of Nestorian Christianity in Central Asia, including a detailed bibliography, is given by N. Sims-Williams in *Encyclopedia Iranica*, vol. 5, ed. E. Yarshater, Mazda, Costa Mesa, California, 1992, pp. 530–35.

35 Professor Michael Zand (Hebrew University, Jerusalem), a leading authority on the history of the Bukharan Jews, has written widely on the subject in Hebrew. Some of the above material was presented in a series of papers, as yet unpublished, delivered at the School of Oriental and African Studies, University of London, 1994–95. See also O. Sukhareva, *Bukhara: XIX – nachalo XX v.*, Nauka, Moscow, 1966, pp. 165–78.

36 Hinduism, though not as strongly represented as Buddhism, was nevertheless influential in the southern Himalayas and in the Tarim basin, particularly in Khotan. One of the best known of the Khotanese Saka texts is a translation of the Ramayana.

37 There are several sections on early Buddhism in north India and Central Asia in *Unesco History*, vol. 2; see also relevant entries in *Cambridge Iran*, esp. vol 2.

38 On the conversion of the Mongols to Buddhism and the move towards a theocracy, see C. R. Bawden, *The Modern History of Mongolia*, Kegan Paul International, London, 1989 (2nd edition), pp. 26–38.

39 S. Akiner, 'Islam, the State and Ethnicity in Central Asia in Historical Perspective', *Religion, State and Society*, vol. 24, nos. 2/3, 1996 (hereafter 'Islam'), esp. pp. 107–116, discusses the fate of Islam under Soviet rule in Central Asia.

40 B. Siklos, 'Mongolian Buddhism: A Defensive Account', in *Mongolia*, esp. pp. 168–82.

41 See, for example, R. D. Schwartz, 'The Anti-Splittist Campaign and Tibetan Political Consciousness', and H. Havnevik, 'The Role of Nuns in Contemporary Tibet', both in *Tibet,* pp. 207–37, and pp. 259–66 respectively.

42 Ole Bruun and Ole Odgaard, 'A Society and Economy in Transition', in *Mongolia in Transition: Old Pattens, New Challenges*, eds O. Bruun and O. Odgaard, Curzon Press, London, 1996, esp. pp.33–37; also S. Akiner, 'Islam', pp. 116–32.

43 *Unesco History*, vol. 2, esp. pp. 405–08.

44 *Unesco History*, vol. 2, pp. 422–33. The latest known examples of the Bactrian script, found in the Tochi valley, Pakistan, date from the 9th century AD.

45 Large finds of documents in Kharosthi-script Gandhari Prakrit have been made on the edge of the Lop Nor desert, at Loulan and Niya.

46 For a discussion of the way in which Brahmi was adapted to fit the needs of these languages see D. Maue, 'A Tentative Stemma of the Varieties of Brahmi Script along the Northern Silk Road', in *Languages and Scripts of Central Asia*, eds S. Akiner and N. Sims-Williams, School of Oriental and African Studies, University of London, London, 1997 (hereafter *Languages and Scripts*), pp. 1–15.

47 W. Sundermann, 'The Manichean Texts in Languages and Scripts of Central Asia', in *Languages and Scripts*, pp. 39–45, discusses the problems of translating and transcribing sacred canonical texts from the original language and script in which they were recorded into other mediums.

48 Talat Tekin, *A Grammar Of Orkhon Turkic*, Indiana University Publications, Bloomington, 1968, pp. 7–30, gives a history of the deciphering of the script and a table of characters; English translations of the main texts are given pp. 261–95.

49 The Arabic script (28 basic symbols) was frequently adapted to the needs of the local languages, often evolving with time as reforms and refinements were introduced. For example, the alphabet that was used from 1865–1923 for Chagatai (also called Old Uzbek), the main Turkic literary language of Transoxiana, had 38 symbols and combinations. The alphabet used for modern Pashto is still developing and now has over 40 letters (on the evolution of this alphabet see D. N. MacKenzie, 'The Development of the Pashto Script', *Languages*, pp. 137–43).

50 In 1983, for example, for a population of some 5 million Kazakhs, 165 newspapers and 32 journals (total annual circulation of 317 million and 34.5 million respectively) were published in Kazakh in Kazakhstan; the number of books (total print run of 12.6 million) were also published. The volume of published material varied somewhat from year to year, but in general the trend was upwards. The situation was similar in the other Central Asian republics. (Annual statistics provided in *Yezhegodnik knigi* and *Letopis' periodicheskikh izdaniy SSSR*, both published by Kniga, Moscow; summaries of entries for relevant languages in S. Akiner, *Islamic Peoples of the Soviet Union*, Kegan Paul International, London, 1986, 2nd ed.)

51 For developments regarding Uighur in the PRC, see Ildiko Bellér-Hann, 'Script Changes in Xinjiang', in *Cultural Change and Continuity in Central Asia*, ed. S. Akiner, Kegan Paul International, London, 1991, pp. 71–83.

52 The most comprehensive study in English of Kazakhstan in this period is G. J. Demko, *The Russian Colonization of Kazakhstan 1896–1916*,

Indiana University Publications, Bloomington, Uralic and Altaic Series vol. 99, 1960.

53 For a more detailed discussion of the role of the nomad heritage in shaping modern identities, see S. Akiner, *The Formation of Kazakh Identity*, Royal Institute of International Affairs, London, 1995.

54 I. M. Franck and D. M. Brownstone, *The Silk Road: A History*, Facts on File Publication, New York, 1986, provides a useful chronological guide to the routes, topography, political setting and economic forces that influenced trade along the Silk Road.

55 Three routes leading to Tibet from the Tarim basin are described in *Ḥudūd al-'Ālam* ('The Regions of the World'), an anonymous geographical work compiled AD 982–3 (E. J. Gibb Memorial, Luzac & Co., 1937, London, translation and commentary by V. Minorsky; p. 61 for text, pp. 254–63 for commentary and translation).

56 Projects to re-establish long-distance trade and transport flows are attracting international backing. See, for example, the current TRACECA project, supported by the European Union, which seeks to re-establish transport networks linking Trans-Caucasia and Central Asia to Europe.

57 For an overview of the environmental problems caused by Soviet development policies see Boris Z. Rumer, *Soviet Central Asia: "A tragic experiment"*, Unwin Hyman, Boston, 1989.

58 In Kyrgyzstan, for example, the number of sheep and goats increased from 2.5 million in 1941, to 10 million in 1991 (see *Narodnoje Khozjajstvo SSSR*, for relevant years).

59 The grasslands in Inner Mongolia formerly provided rich pasture and hay-cutting meadows; as a result of the campaigns of 1958–62, and 1966–73, 1,200,000 hectares of grassland have been desertified (*Culture and Environment in Inner Asia: the Pastoral Economy and the Environment*, eds C. Humphrey and D. Sneath, White Horse Press, Cambridge, 1996, p. 112).

60 Melvyn C. Goldstein, 'Change, Conflict and Continuity among a Community of Nomadic Pastoralists: A Case Study from Western Tibet, 1950–1990', in *Tibet*, pp. 76–90, contrasts the situation in the TAR with that in other parts of 'historic' Tibet; he gives a clear exposition of the issues relating to the definition of Tibet, setting out the arguments for distinguishing between developments in 'the political entity of Tibet' and those in 'the ethnic border areas' – an important distinction that is often overlooked.

61 The Kazakh population numbered some 4,120,000 in 1930; by 1940, it had been reduced by over 40% as a result of starvation, epidemics and executions. The livestock losses in the period 1928–35 were also devastating: from 6.5 million to under 1 million head of cattle; 18.5 million to 1.5 million sheep; 3.5 million to under 500,000 horses; 1 million to 63,000 camels. (See further Zh. Abylkhozhin, *Traditsionnaya struktura Kazakhstana*, Gylym, Alma-Ata, 1991, pp. 184–90; M. B. Tatimov, *Sotsial'naya obuslovlennost' demograficheskikh protsessov*, Nauka, Alma-Ata, 1989, pp. 120–26.)

Selected Bibliography

Akiner, S. (1986), *Islamic Peoples of the Soviet Union,* Kegan Paul International (2nd ed.), London.

Bregel, Y. (comp. and ed.) (1995), *Bibliography of Islamic Central Asia,* 3 vols, Indiana University Publications, Uralic and Altaic Series, Bloomington.

Dani, A. H., Masson, V. M., *et al* (1992–98), *History of Civilizations of Central Asia,* vols 1–3, Unesco Publishing, Paris.

Franck, I. M., and Brownstone, D. M. (1986), *The Silk Road: A History,* Facts on File Publications, New York.

Krader, L. (1971), *Peoples of Central Asia,* Indiana University Publications, Uralic and Altaic Series, Bloomington.

Legg, S. (1971), *The Heartland,* Farrar, Straus & Giroux, New York, 1971.

Sinor, D. (ed) (1990), *Cambridge History of Early Inner Asia,* Cambridge University Press, Cambridge, 1990.

PART TWO

THEORETICAL APPROACHES TO SUSTAINABLE DEVELOPMENT

Chapter 2

Development and Globalisation
Social, Psychological and Environmental Costs

Helena Norberg-Hodge

Virtually every political leader today – in both the North and the South – believes that the goals of society are best met if the economy keeps growing. Most support the notion that continued growth can be achieved by internationalising the economy and liberalising trade. Around the world, this internationalisation – which means tying ever more people and cultures into a single, centralised world economic system – is seen as a means of reviving ailing, bankrupt economies.

From a social and ecological point of view, however, this is a disastrous course. Rather than curing present problems, the globalisation of the economy will accelerate social and environmental breakdown. It will mean a dramatic increase in environmental pollution, a further erosion of family and community ties, and a marked weakening of real democracy. Instead, strengthening and diversifying local, regional and national economies should be high on the agenda of developed and developing countries alike.

My perspective is based on studying many societies at different levels of industrialisation. This has revealed the extent to which the complex interactions between technology and economics inherent in the industrial model affect not only the environment, but our culture and world view as well. I have lived and studied in America, where industrialisation has proceeded furthest; in Sweden, with its socialist variant of industrial society; in Innsbruck, Austria, 25 years ago, when it was relatively uninfluenced by economic development; and in rural Spain, where European Union trading policies have had a profound impact on rural life. Most importantly, I have had the rare opportunity over the past two decades to observe firsthand the effects of development on an ancient traditional culture: Ladakh, or 'Little Tibet', in Jammu and Kashmir state, India. The changes that have occured in Ladakh in this period are roughly equivalent to those that occurred over a period of 500 years or so in many other cultures. The situation in Ladakh can thus help to illustrate the impact of the West – from the conquistadors to colonialism to development – on other cultures. The lessons of Ladakh are relevant even in the most industrialised parts of the world, but are particularly so in Central Asia, where the situation in many countries is similar to that of Ladakh.

Twenty years ago, life in Ladakhi villages rested on the same foundations as it had for centuries. The economy was primarily based on subsistence farming, with trade limited to luxuries. Cultural practices and an awareness of the limits of resources kept population within the bounds that the land could support. There was virtually no waste or pollution. Within the decentralised village structure the household was the centre of economic activity, and the status of women was high; in many cases they were the decision-makers. The old remained respected and productive members of society through-out their lives. Crime, unemployment and homelessness were essentially unknown. Thanks to an intimate, location-specific knowl-edge of their ecosystem, the Ladakhis had prospered, providing all their material needs – as well as beautiful works of art, drama and music – while having far more time for friends and leisure activities than the average Westerner.

The traditional way of life in Ladakh was certainly not perfect. However, it is clear that the changes wrought by development have not, on balance, been an improvement. As Ladakh is steadily tied into the global economy, the gap between rich and poor is widening; waste and pollution are rapidly increasing; the status of women is declining; the old are being forgotten and the young are growing ashamed of their own culture. A once self-reliant people is becoming increasingly dependent on imported resources over which they have no control, and relying ever more upon a distant bureaucracy for needs once provided by their own community. By almost any meaningful measure, the quality of life is declining. And yet because of a reliance on narrow economic yardsticks, the destructive changes occuring in Ladakh are generally seen as mere side-effects of the positive transformation called 'development'.

The promise of conventional development is that by following in the footsteps of the 'developed' countries of the world, the 'under-developed' countries can become rich and comfortable too. Poverty will be eliminated, and the problems of overpopulation and environmental degradation will be solved. This argument contains an inherent flaw, even deception. The fact is that the developed nations are consuming essential resources in such a way and at such a rate that it is impossible for the 'underdeveloped' areas of the world to follow in their footsteps. When one quarter of the world's population consumes three quarters of the world's resources, and then in effect turns around and tells the others to do as they do, it is little short of a hoax. Development is all too often a euphemism for exploitation, a new colonialism. The forces of development and modernisation have pulled most people away from a sure subsistence and got them to chase after an illusion, only to fall flat on their faces, materially impoverished and psychologically disoriented. A majority are turned into slum dwellers, having left the land and their local economy to end up in the shadow of an urban dream that can never be realised.

Development and the corporate-led trend toward globalisation depend upon continuous government investments. They require the building-up of a large-scale industrial infrastructure, including roads, mass communications facilities, energy installations, and schools for specialised education. Among other things, this heavily subsidised infrastructure allows goods produced on a large scale and transported long distances to be sold at artificially low prices, in many cases at lower prices than goods produced locally. In Ladakh, the Indian government is not only subsidising roads, schools and energy installations, it is also bringing in subsidised food from India's breadbasket, the Punjab. The food arriving in lorries by the tonne is cheaper in the local bazaar than food grown five minutes' walk away. Although Ladakh's local economy has provided enough food for its people for 2,000 years, many Ladakhis no longer find it worthwhile to continue farming.

This same process is affecting a whole range of products, from clothes to household utensils to building materials. Goods imported to Ladakh from distant parts of India can often be produced and distributed at prices far lower than goods produced locally: again, because of a heavily subsidised industrial infrastructure. The end result of all this long-distance transport of subsidised goods is that Ladakh's local economy is being steadily dismantled, and with it the local community that was once tied together by bonds of inter-dependence.

This trend is exacerbated by other changes that have accompanied economic development. Traditionally, children learned how to farm from relatives and neighbours; now they are put into Western-style schools that prepare them for specialised jobs in an industrial economy. In Ladakh, these jobs are very few and far between. As more and more people are pulled off the land, the number of unemployed Ladakhis competing with each other for these scarce jobs is growing exponentially. What's more, the course of the economy, once controlled locally, is increasingly dominated by distant market forces and anonymous bureaucracies. The result has been a growing insecurity and competitiveness – even leading to ethnic conflict – amongst a once secure and cooperative people. A range of related social problems has appeared almost overnight, including crime, family break-up and homelessness. And as the Ladakhis have become separated from the land, their awareness of the limits of local resources has dimmed. Pollution is on the increase, and the population is growing at unsustainable rates.

Economists would dismiss these negative impacts, which are not so easily quantifiable as the monetary transactions that are the goal of economic development. Yet the situation in Ladakh vividly illustrates the shortcomings of defining human welfare in terms of economic growth and its main indicator, gross domestic product (GDP). In traditional parts of Ladakh, the standard of living is actually quite high when compared with most of the Third World. Yet because people provide for almost all their needs outside the cash economy, the GDP is virtually zero; Ladakh appears to be at the very bottom of the economic order, in dire need of economic development. In effect this means that no distinction is made between self-sufficient Ladakhi farmers and the homeless on the streets of New York. In both cases there may be no income, but the reality behind the statistics is as different as day and night.

In Ladakh and elsewhere in the Third World, farmers once grew a variety of crops and kept a few animals to provide for themselves – either directly or through the local economy. Now, development is encouraging these farmers to grow cash crops for distant markets. As a result, they are becoming dependent on forces beyond their control: huge transportation networks, oil and agrochemical markets, the fluctuations of international finance. Over the course of time, inflation obliges them to produce more and more to secure the income that they now need in order to buy what they used to grow themselves. Throughout the world, development has thus displaced and marginalised self-reliant local economies in general, and small farmers in particular.

I should point out that it is not only the South that is affected by this process. The pressures of economic and technological change are bearing down even on the industrialised societies in a similar way; we too are being 'developed'. Today, even though only 2 or 3% of the population is left on the land, small farmers are still being squeezed out of existence; and even though industrialisation has pared the extended family down to a small nuclear unit, our economy is still chipping away at it. Technological advance is continuing to speed life up, while robbing people of time. Increased trade and ever greater mobility are furthering anonymity and a breakdown of community. In the West, these trends are labelled 'progress' rather than development, but they emanate from the same process of industrialisation that inevitably leads to centralisation, social degradation, and the wasteful use of resources.

'Progress' based on economic and technological development has reached an advanced stage in the West. Wherever we look, we can see its inexorable logic at work: people are replaced with machines, local interdependence with global markets, country lanes with freeways, and the corner shop with a supermarket. Governments around the world, regardless of their political hue, are encouraging an *acceleration* in these changes through support for a globalised economy. Long-distance trade receives heavy subsidies, in particular, to maintain and expand networks of communication and transportation. Meanwhile, trade agreements justified by the rhetoric of 'free trade', like the General Agreement on Tariffs and Trade, the North American Free Trade Area and the Maastricht Treaty in the European Union, are leading to ever greater centralisation of economic and political power. Local communities are rapidly losing control over their own destinies, and even national governments are handing over control to supranational institutions such as the European Union, World Bank, and now the World Trade Organisation. Such organisations are so far removed from the people they are supposed to represent that they are incapable of responding to their diverse interests.

Even though the phrase 'think globally, act locally' is heard frequently these days, the thrust of development and progress is entirely in the direction of globalisation. Local cultures and economies are rapidly disappearing, taking animal and plant species with them. In the process, our very existence is threatened.

In the natural world, diversity is an inescapable fact of life. We are just beginning to discover how important even the most 'insignificant' insect or plant can be for our survival. The accelerating pace at which diverse species of life are being eradicated has, in fact, become a major issue. *Cultural* diversity is as important as diversity in the natural world and, in fact, follows directly from it. Traditional cultures mirrored their particular environments, deriving their food, clothing, and shelter from local resources. Even in the most developed parts of the world today, there are still remnants of local adaptation to diversity. In the American Southwest, for example, you find flat-roofed adobe houses, ideally suited to the extremes of the desert climate; in New England, houses are made of locally abundant wood, and have peaked roofs designed to shed the rain and snow. The cuisines of different cultures still reflect local food sources, from the olive oil prevalent in Mediterranean cooking to the oatmeal and kippered herring on the Scotsman's breakfast table.

The emerging world economy and the growing domination of science and technology are breaking down this natural and cultural diversity by remaking societies around the world, and particularly in the South, so as to conform to a single monoculture. Both globalisation and development are based on the assumption that needs are everywhere essentially the same, that everyone can eat the same food, live in the same type of house, wear the same clothes. The same cement buildings, the same films and television programmes find their way to the most remote corners of the world. Across the world, *Dallas* beams into people's homes, and pinstripe suits are *de rigeur*. Lingonberry and pineapple juice are giving way to Coca Cola; woolen robes and cotton saris are being replaced by blue jeans; and yaks and highland cattle are disappearing in favour of Jersey cows. Virtually identical toy shops – in Ladakh, in Beijing, in remote mountain villages in Spain – sell the same blonde, blue-eyed Barbie dolls and Rambos with machine guns. Even language is being homogenised, since English is the *lingua franca* of the modern 'global community'.

The breakdown of cultural diversity throughout the South is accelerated by grossly distorted impressions of modern life. The predominant image is one of ease and glamour: everyone is beautiful, everyone is rich. People see the fast cars, the microwave ovens, and the video machines. Advertising and the media are telling them what to do, in fact telling them what to *be*: modern, civilised and rich. What isn't seen are the side-effects of this way of life: the environmental deterioration, the psychological stress, the drug addiction, the home-lessness. People who have been presented with only one side of the development coin are left vulnerable and eager for modernisation.

In this way, diverse cultures from Alaska to Australia are being overrun by the industrial monoculture. This is a tragedy of many dimensions. With the destruction of each culture, we are erasing centuries of accumulated knowledge. And as diverse ethnic groups feel their identity threatened, conflict and social breakdown almost inevitably follow.

Today, these trends are supported by government spending, a large portion of which is devoted to expanding transport and communication infrastructures, enlarging global markets and supporting the demands of large-scale producers. Throughout the developing world, money flows freely into huge projects aimed at increasing market transactions and boosting GDP. Heavy subsidies are available for massive dams, nuclear energy installations, television and radio facilities, highways, fertiliser plants and airports, all of which serve to reinforce dependence on centralised systems. The pattern is the same in the industrialised nations as well. Government and industry continue to push towards a larger scale, greater centralisation, and further dependence on the multinational corporations which produce and deliver the goods that are traded world-wide.

Reversing these trends and solving today's social and environmental crises will require active steps towards decentralisation. This does not mean retreating into cultural or economic isolationism, but rather reducing the influence of powerful global economic forces so as to allow each region to nourish its own economy and its own traditions. Since extreme dependence has already been created on both national and international levels, it would be irresponsible to 'delink' economies from one day to the next. Countries in the North cannot, for example, suddenly halt their purchase of coffee, cotton or manufactured goods from those countries in the South whose economies totally depend on such trade. But they *can* immediately begin supporting aid programmes that will enable people to provide basic needs for themselves; ensuring, for instance, that the production of food for local consumption has priority over exports to the West. Small and medium producers the world over would be better off as a result. They would also benefit if their products did not have to compete with those shipped great distances via subsidised transport networks, and if support were given to developing technologies appropriate for local conditions, rather than labour-displacing, capital-intensive equipment suited to large-scale industry and agribusiness.

One of the most effective ways of supporting local economies would be to reduce unnecessary trade. We are currently transporting across whole continents and several oceans a vast range of products, from milk to apples to furniture, that could just as easily be produced in their place of destination. By reducing and eliminating subsidies for transportation, we would cut waste and pollution, improve the position of small producers, and strengthen local economies and communities in one fell swoop.

What exactly is 'local,' and what is 'necessary' as opposed to 'unnecessary' trade are issues that cannot be defined in absolute terms. But the crucial point is that the *principle* of heavily subsidised international trade is one that needs critical reassessment. The goal would not be to encourage narrow protectionism, but rather to allow for the sustainable and equitable use of natural resources worldwide. It is in robust, local-scale economies that we find genuinely 'free' markets: free of the corporate manipulation, hidden subsidies, waste, and immense promotional costs that characterise today's global market.

Parallel to economic decentralisation, the production of energy would need to be decentralised. This would be a positive step in both the North and the South, but because the energy infrastructure of most developing countries is still relatively limited, the widespread application of renewable energy technologies in these regions would be comparatively easy. Local communities and economies there would be better off if real support were given to small-scale projects based on locally available resources, such as village-scale hydroelectric installations, or solar ovens and water heaters for the household. Until now, however, such support has simply not materialised. Instead, the West has pushed its own industrial model, based on large-scale, centralised power production. One of the greatest scandals of development is that despite tremendous potential, not a single country in the developing world has managed to promote small-scale decentralised applications of solar energy on anything more than a token basis.

If the diverse cultures of the South are to survive the pressures leading towards global monoculturisation, another urgent requirement is a public information campaign to correct the incomplete and misleading images of the industrialised world. Information about the long-term effects of everything from powdered milk for babies to a dependence on fossil fuels tends not to reach the least developed areas of the world. And the seductive images in the media and advertising that do arrive are not accompanied by warnings about toxic wastes, the erosion of farmlands, acid rain, or global warming. Rather than more development, what is needed is what I call 'counter-development'.

72

The primary goal of counter-development would be to provide people with the means to make fully informed choices about their own future. Using every possible form of communication, from satellite television to storytelling, it should be publicised that today's capital- and energy-intensive trends are simply unsustainable. Ultimately, the aim would be to promote self-respect and self-reliance, thereby protecting life-sustaining diversity and creating the conditions for locally based, truly sustainable development.

At the same time, counter-development would promote and popularise a new, wider, and more humane definition of progress. It would highlight some of the innumerable local initiatives in the North that are exploring more sustainable alternatives. It would point to the viability of traditional systems as well as bringing information about new trends in agriculture: about permaculture, biodynamics, and the booming movement toward organic methods of cultivation. It would report on bioregionalism and local economic systems; it would publicise the windmills in Denmark and California, and the growing demand for acupuncture, homeopathy, and other nature-based systems of health care. It would make more visible the enormous interest around the world in environmental protection, soil conservation, and air and water quality.

In the more industrialised parts of the world, the distance between producers and consumers has now grown enormous, and strengthening local economies means *shortening* those links. This could happen if the government spending and subsidies that now go to support the further globalisation of the economy were shifted to support instead the needs of more localised economic activities. This would mean giving support to the many grassroots initiatives that are already springing up throughout the world, ranging from local banks and currencies, to barter and local employment trading systems, to more direct links between farmers and consumers. Such measures enable communities and towns to reduce unemployment, cut back on pollution and waste, and enhance their connection with the natural world.

The stronger sense of community that stems from shorter producer-consumer links has important psychological benefits in turn. Recent research has made it clear that the rise in crime, violence, depression, even divorce, is to a very great extent a consequence of the breakdown of community. Conversely, children growing up with a sense of connection to their place on the earth and to others around them – in other words, children who are embedded in a community – grow up with a stronger sense of self-esteem and healthier identities, thereby helping to solve many of today's most pressing social problems.

Greater self-reliance in the North would reduce the pressures on the South to focus on export markets rather than production for local consumption. Despite the free-trade rhetoric, it is not in the long-term interests of the majority of people in the South to sell their labour and resources at cut-rate prices to the corporate interests that market them to the populations of the North.

The momentum of development and progress, supported by massive investments from government and industry, are currently propelling the planet towards a global monoculture. The social and environmental consequences of this trend are potentially disastrous. Today's policies put technology and the needs of the economy at the top of the agenda. A significant shift in direction, leading to a better balance between local economies and global trade, is clearly needed. It is hoped that the peoples of Central Asia can help to lead the way.

Chapter 3

The Shortcomings of the Classical Economic Model

Appropriate Economic Parameters are required for the Sustainable Development of Central Asia

Sander G. Tideman

Introduction: Capitalist Economics as the Only Viable Economic Ideology?

Free market capitalist economics is a very powerful set of ideas, now very much in vogue in Central Asia. When its main rival, communism or state-planned socialist economics, was discredited and abandoned in the Soviet Union and Eastern Europe, and to a large extent also in China, Western-style free market economics became the only viable economic theory available for governments to run and organise their societies. In fact, because of the stunning collapse of communist-inspired regimes and the economic 'miracles' of the Asian Pacific regions including China, most authorities perceive free market capitalism to be superior to communist economic theory, and have embraced the new set of ideas with much vigour. The laws of present-day capitalist economics are so appealing and pervasive that many of us largely take them for granted, like the laws of motion and gravity. Indeed, the principles underlying capitalist economics were codified by Adam Smith on the basis of the laws of physics devised by Sir Isaac Newton at the beginning of the Scientific Revolution – and have not changed much in the last two centuries in spite of significant developments in our societies and the laws of physics.[1]

While capitalist economic values have been eagerly adopted by formerly socialist regimes, in the West a new school has emerged that seeks to address the shortcomings of the classical economy as it is practised – mostly driven by concern for the environment. The high costs of the classical growth-oriented economic model, from environmental degradation to cultural and community disruption, which are particularly evident in developing countries, are now quite obvious. In this paradigm shift, a debate has emerged over old paradigm contradictions and anomalies; for example, what to measure and how to measure it; and how to treat those values and resources that have no price.

Although there is no alternative comprehensive economic model with any credibility in the world at present, it might be a useful exercise to look more closely at the classical economic theory, and the way it is currently practised, and link this to the prospects for sustainable development in Central Asia. For the indiscriminate and overly enthusiastic adoption of a system that is increasingly being recognised as partially 'blind' could bring irreparable harm to the fragile social and natural environment of Central Asia.

The Shortcomings of the Classical Economic Theory

The only values which appear in classical economic models are those that can be quantified by being assigned monetary weightings. This emphasis on quantification gives economics the appearance of an exact science. However, it severely restricts the scope of economic theories by excluding qualitative distinctions that are crucial to understanding the ecological, social, and psychological dimensions of economic activity. For example, economic calculations ignore the value of things that cannot be easily quantified, such as fresh water, green pastures, clean air, the beauty of mountains, traditional rural ways of life, to name but a few. In fact, this partial 'blindness' of our current economic system is recognised as the most important force behind the accelerating destruction of the global environment.[2]

Consider the most basic measure of a nation's economic performance, gross national product (GNP), which governments want to see grow each year. In calculating GNP, natural resources are not depreciated as they are being exploited. Buildings and factories are depreciated, as well as machinery, equipment, trucks and cars. So why, to take an example that is familiar for Central Asians, is pastoral land not depreciated when it turns into desert after irresponsible grazing and farming methods have reduced its ability to resist wind, rain and frost? Why is that loss not measured as an economic loss of the process to produce meat and milk from cattle? If the destruction of the grasslands is high enough in a given year, in actuality the nation may end up poorer, even if the value of animal products is taken into account. At the same time, the national economic statistics will show that the country has grown richer for having raised the cattle and traded animal products, without taking into account the money that should be spent to protect the pastures from desertification.

There are hundreds of other examples. A Tibetan saying in eastern Sichuan goes: 'The forested mountains are our bank, the trees are the bank's interest'.[3] For generations this saying reflected wisdom, for indeed the number of trees was practically limitless and the demands placed on the environment by the local community were extremely modest. In the last decade, however, this Tibetan wisdom has degenerated into ignorance, fuelling environmental destruction – simply because it has been exposed to national policies which are based on the flawed logic of capitalist economics. The money received from the sale of logs is counted as part of that region's income for the year. The funds spent on the chain-saws and logging trucks will be entered on the expense side of the accounts, but those to be spent on the supposed replanting of trees will not. In fact, nowhere in the calculations of this country's GNP will be an entry reflecting the distressing reality that millions of trees are gone, many of them forever.

Classical economic theory holds that all participants in the market between supply and demand have 'perfect information' about the facts surrounding and supporting their choices. This is another assumption that has proven to be incorrect, especially in light of the inability of classical economics to account for lost natural resources. Our current economic system not only makes unrealistic assumptions about the information available to real people in the real world; it also assumes incorrectly that natural resources are limitless 'free goods', failing to distinguish between renewable and non-renewable goods, and simply equating them on the basis of monetary values set by the supposedly 'informed' market.

Similarly, classical economic theory measures the efficiency of production, or 'productivity', in a way that keeps track of the good things we produce but not the bad. For example, if a country produces coal, its inevitable by-product, coal dust, is not accounted for. Later, when funds are required to clean up the pollution that the coal mines created, they are normally included in the national accounts as another positive entry on the ledger. In other words, productivity is defined narrowly and encourages us to equate gains in productivity with economic progress – a definition traditionally applied in capitalist and communist systems alike.

Our economic system also fails to account for all the associated costs of what is called consumption. Every time we consume something, some sort of waste is created, but this fact is overlooked by classical economic textbooks. For example, for all the coal and oil we consume in a given day, we do not account for the extra carbon dioxide (CO_2) in the atmosphere. Since in the classical theory an increase in 'standard of living' is equated with an increase in consumption, this misconception is encouraging us to produce more and more, and thus also more waste, regardless of the environmental consequences – again a misconception that has taken root all over the world.

Classical economic theory also contains questionable assumptions about what is valuable in the future as opposed to the present. In particular the standard discount rate that assesses cash-flows resulting from the use or development of natural resources assumes that all resources belong totally to the present generation. As a result, any value that they may have to future generations is heavily 'discounted' when compared to the value of using them up now. Thus, the dominant economic ideology guiding our economic behaviour discriminates against future generations. Simply because it seems impossible to put a price on the environmental effects of our economic choices, we tend to treat natural resources as worthless.

We measure economic performance in exactly the same way the world over, in both developed and developing countries, and as a result we tend to ignore the social and environmental costs of development. The classical method of calculating GNP not only distorts the reality of how our economic behaviour affects the environment, it also leads to the implementation of well-intentioned policies that ultimately turn out to be harmful for both nature and society. For example, when multilateral development banks decide what kind of loans to give to developing countries, most of them base their decisions on how these loans might improve the recipients' economic performance, which is invariably measured by the movement of GNP. Yet there is a growing concern that GNP growth policies – both in the South and the North – widen poverty gaps and exacerbate unemployment. The *Human Development Report 1993* of the United Nations Development Programme indicates that jobless economic growth is increasing around the world. The OECD (Organisation for Economic Cooperation and Development) countries face new dilemmas: creeping budget deficits and jobless growth are the symptoms, but old remedies no longer apply.[4]

It should, of course, be quite obvious that unlimited expansion in a finite environment leads to disaster, but our economic system continues to be preoccupied with growth. By concentrating on the mere statistics of GNP, we fail to distinguish between the qualitative aspects of growth: healthy or unhealthy growth, temporary or sustainable growth. We do not question what growth is actually needed, what is required to actually improve the quality of life. In contrast, the King of Bhutan, a secluded Buddhist kingdom in Central Asia, has been speaking of measuring his country's development by 'GNH', Gross National Happiness, to stress that the definition of growth should incorporate more qualitative indicators.

The Shortcomings Can Be Amended, but With Difficulty

It should be said that in the last few decades, environmental economists have been seeking to amend the shortcomings in economic theory, and several improvements have been made.[5] For example, UNDP has since 1990 published the Human Development Index (HDI), which ranks countries by a 'qualitative' measurement that combines life expectancy, educational attainment and basic purchasing power. Likewise, the OECD is producing useful work on indicators which not only incorporate expenditures for basic social factors such as health and education, but also the value of environmental resources. The application of these indicators would indeed be a major advancement in government accounting.

However, very few of these advances in economic theory have been implemented in the practice of governments and policy-makers, mainly because of the difficulty of agreeing on the value of these resources. So, in all fairness, we should differentiate between classical economic theory, which is gradually being improved, and the economic ideology generally adhered to by policy-makers and the general public.

In fact, many of our current economic policies that can be faulted for environmental degradation go against the spirit of free market economics. For example, many countries have subsidised living costs in urban areas by controlling food prices and providing cheap housing, schooling, transport and telecommunications facilities, while wages in the formal and largely urban industrialised sector have been maintained at artificially high rates. These subsidies have favoured life in the cities over that in the rural areas, leading, amongst other consequences, to migration from rural areas into the cities, most notably in developing countries.

Similarly, the theory of free market economics would recommend allowing the choice of capital- or labour-intensive technologies to be determined by market prices. Since in developing countries where capital is scarce, the free market price of capital is high, companies are expected to adopt labour-intensive technologies. Yet this logic has been contradicted in many countries by deliberate government policies to reduce the cost of capital (through artificially low interest rates) and increase the cost of labour (by insisting on full benefits in the formal state sector), specifically because adopting capital-intensive technologies was believed to be a sound development strategy. These distortions of free market economics are a result of political choices, made mostly by urban elites who generally dominate national governments and who are inclined to organise the economy in their own favour, and not of economic theory as such.

In other words, in order for us to solve the world's present social and environmental problems, we should not only seek to improve the economic theory, but also carefully examine the popular beliefs and values on the basis of which we develop and implement economic policies in our societies. Until we have rectified the conceptual flaws and short-sighted assumptions underlying our current economic model, and have devised a more environmentally friendly and appropriate economic theory, Central Asian authorities are well advised to have a second look at the economic model they have adopted.[6]

The Proper Use of Central Asia's Land

Capitalist economic theory considers consumption to be the sole end and purpose of all economic activity, taking production factors like land, labour and capital as the means. Countries in Central Asia tend to lack large reserves of capital and labour, while much of their huge landmass is unsuitable for production because of long distances, geographical barriers and an extreme, dry climate. Nonetheless, out of mere necessity, land – and the mineral resources it contains – is likely to be treated as their main or only means of production. This is a question which should be considered carefully.

First, we have to look at simple economic factors. In Western economies the importance of land-based economic activities has decreased with industrialisation. But in Central Asia, with its abundance of land and shortage of other means of production, land-based activities such as agriculture and nomadic pastoralism will remain a dominant if not substantial part of the economy. Even if Central Asia opens completely to the industrialised global economy, the economies of Central Asia will continue to depend to a large extent on contributions from the land, which makes them essentially different from the economies of the Western industrialised world.

Yet few of the national economic planners seem to realise the fundamental differences between Central Asia's land-based economies and the classical capitalist model that arose from the industrial experience in the sedentary areas of colonial Europe. China's recent emphasis on monetary growth, for example, which is inspired by classical economic principles, has had an immediate negative impact on the lands in western China. The local authorities of Qinghai Province, which contains much of China's pastoral lands, have been told by Beijing to significantly raise economic yields, while subsidies from higher authorities have been reduced or halted. As a direct consequence, taxes have been imposed on nomadic herders who have thus far largely remained outside the money economy. In order to regulate ownership and thus facilitate tax-collection, pastoral lands have been fenced, inviting problems such as overherding and overgrazing – and the consequential destruction of the soil.[7]

Through processes like these, which clearly reflect lack of understanding of nomadic life in government circles, civilisations based on land are irreversibly changed. Central Asians, therefore, need to ask themselves a more fundamental question. In societies that are traditionally nomadic or semi-nomadic, is the land merely a means of production or is it something more, something that is an end in itself? It is no use to answer this question economically or scientifically. It is a political, or even philosophical question.

In modern times, the main danger to the soil, and therewith not only to agriculture, or pastoralism, but also to the civilisation based on the soil, stems from the tendency to exploit land simply as a means of production. Western economics have applied to agriculture the principles of industry, without asking whether it might be something essentially different. The ideal of industry is to produce in the cheapest and most efficient way. But with economic growth, labour tends to become more expensive, and thus industry switches to labour-saving technologies. As a result, human activities, including farming and herding, are minimised and the productive process is gradually turned to high-tech machines.

Many native Central Asians, however, still regard agriculture, or pastoralism for that matter, as a primary human activity, whereas for them industry is a derived, or secondary, activity. The latter relies on the former. In this view, human life and traditional society can continue without industry, whereas it cannot continue without agriculture. This viewpoint, incidentally, is not unknown to the West; many European governments are still locked in a dispute about the protection of their respective agricultural industries, simply because in the final analysis the farming community means more to the national identity than was initially thought.

One can argue that political or philosophical considerations do not belong to the realm of economics. And even if they do, how on earth does one measure the 'qualitative' value of the land? But the alternative to not answering this question is clear: if monetary values are the ultimate criteria and determinants of human action, the value of land and the civilisation based on that land will be minimal, and ultimately destroyed. Thus, the great nomadic civilisations such as the Kazakhs, Mongols and Tibetans are likely to survive only in designated protected parks and reserves, more as an historical curiosity than as a viable way of living – a fate not unlike that of Native American culture in North America.

New Members of the World Market

The economic success of the export-oriented economies of East Asia and the evidence of the shortcomings of inward-oriented industria-lisation strategies of the past, have led Central Asian governments to be preoccupied with generating exports and attracting foreign investment. The rationale behind outward-oriented growth is to maximise economic well-being by pursuing comparative advantage vis-à-vis foreign countries. All Central Asian states are pinning their hopes on their abundant mineral and energy sources. Turkmenistan has the brightest prospects because of its abundant natural gas, followed by Kazakhstan which is believed to hold considerable untapped oil reserves. But the economic value of most of these resources is questionable. Due to high exploitation costs they are likely to remain untapped for many years to come. Curiously, most of the economic success stories of recent decades have been resource-poor countries, such as Taiwan and South Korea, while richly endowed countries such as Zaire and Zambia have performed below expectations.

The perceived need to export is aggravated since Central Asian countries have small markets with little purchasing power. Obviously, if a country produces for export to the world market (mostly comprising richer countries), it can take the availability of purchasing power for granted. Further, exports generate foreign exchange, which is regarded as infinitely more valuable than local currency, since the latter, often because of high inflation rates, is subjected to devaluations versus hard, foreign currencies.

Therefore, only production for export is believed to be proper development. For land-locked countries that are sparsely populated and have little capital, this reasoning deserves further thought.[8] Consider the following facts. As the supply of labour in Central Asian countries is not abundant, labour-intensive technologies are not so much of an option as in the cheap-labour economies of East and South Asia. These countries, especially the 'Tigers' of the Asian Pacific, have attracted massive foreign investment primarily because of the promise of low-cost labour. Thus in order for the Central Asians to compete in world markets, they will be required to employ the capital-intensive technology of the rich countries. This requirement leads to spending a large proportion of foreign exchange reserves on imports, or on the repayment of debt incurred to purchase the technology.

Many potential foreign investors, on the other hand, who could assist in transferring the required capital and technology to these countries, are deterred from investing substantially in Central Asian countries by the limited local markets and the absence of a large pool of cheap labour. Add to this the fragility of the land, and it is clear that Central Asia is not well placed to compete in world markets with countries such as those on the Asian Pacific coast.[9]

More importantly, it needs to be considered that the average people of Central Asia do not live by exporting; for most of them it is a very alien thing to do, and what they produce for themselves and for each other is of much greater importance to them than what they produce for a foreign market. Again, this is a matter which requires qualitative judgement: to what extent does export improve the quality of people's lives?

Export Versus the Local Market: Finding the Right Balance

It cannot be denied that exports and foreign investors are important for development. For one, they allow local industry and authorities to learn from foreign countries and to select appropriate technology from these countries. But at least of equal importance is the development of industry for the local market, even though there may be little purchasing power in that market at present. The fact that in Ulaanbaatar there is a shortage of milk and meat, while Mongolia has more than 25 million head of cattle, illustrates this point quite clearly.[10]

Moreover, the modern preoccupation with exports – and the automatic requirement to maintain fair trading relations enshrined in the principles of the General Agreement on Tariffs and Trade – leaves small Central Asian nations exposed to curious trading patterns. Mongolia is importing bottles of mineral water from Hong Kong (for the richer people in the cities), although this Central Asian country is one of the few remaining regions in the world endowed with huge and unspoiled fresh water reserves. Economic planners should be concerned with the development of local industry, not because of an intuitive distrust of foreign markets, as was the case in the Soviet period, but because it ultimately leads to the substitution of *unnecessary* and expensive imports – and hence the preservation of precious foreign exchange.

Yet most economic experts in Central Asia have tended to look at the (until recently) successful economic models of Taiwan, South Korea, Hong Kong and Singapore and more recently of coastal China, which rely heavily on exports to foreign countries. The development plan for the Tibet Autonomous Region is strongly influenced by the model of Shenzhen, a 'Special Economic Zone' on China's eastern coast which in the last decade has boomed as an export processing zone for nearby Hong Kong. Inspired by this success, the government plans the establishment of a high-tech export zone near Lhasa, Tibet's land-locked capital at 4,000 m altitude, apparently without questioning the appropriateness of such capital- and energy-intensive industry in a remote region that lacks sufficient capital, labour and energy resources. One can find similar plans in other parts of China's Central Asia, for example in Korgas along the Xinjiang-Kazakhstan border.[11]

The coastal provinces of China hold a fascination for many Central Asian policy-makers, especially in Kazakhstan, Kyrgyzstan and Mongolia, which border China and have seen socio-economic ties expand rapidly since 1991. The crucial difference, however, between China and the Central Asian states is the large share of the Chinese labour force employed in agriculture on the eve of economic reform in 1979. Chinese agricultural workers, whose productive energies were released by the household responsibility system, under which they obtained considerable economic freedom, had nothing to lose from the reforms. By contrast, former-Soviet-Union countries, including the

Central Asian states, faced the more difficult task of relocating workers from inefficient state enterprises – workers who feared that they would suffer during the reform process.[12]

Focusing on statistics alone, one easily overlooks the fact that the East Asian growth countries had very different characteristics to start with, such as large reserves of cheap labour, some basic infrastructure (harbours and other transportation facilities), relatively easy access to foreign capital, ties with richer overseas ethnic communities and – above all – an advantageous geographical location.[13] Central Asian nations are land-locked, have few transportation facilities,[14] restrictive physical conditions and little capital and labour, and, most distinctively, have fragile soils that for centuries have proven to be suitable only for nomadic or semi-nomadic forms of society. To copy blindly a model based on export-generation and the availability of capital and large quanties of cheap labour, will be a mistake which Central Asian nations can hardly afford.

Avoiding this mistake requires not only the establishment of local industry with appropriate technology for a local market, but also that economic activity should rely on appropriate resources, such as decentralised, locally available and renewable sources of energy (solar, wind, biomass, geothermal etc.), rather than on centralised, imported and non-renewable energy.[15] The same applies to irrigation. The arid lands of Central Asia need small-scale irrigation facilities which can be managed by local communities who understand the needs and contraints of the land. The dramatic shrinking of the Aral Sea, whose water was used to grow cotton for export, serves as a sad reminder for Central Asia that large-scale centralised systems devised by ambitious central planners can lead to unparalleled environmental and human catastrophes.[16]

Conclusion: Quantity or Quality?

The purpose of this discussion is not to merely criticise the current economic system, for, as I stated, there is no alternative economic model with any credibility at present, and admittedly the classical economic theory has been an extremely powerful tool in raising the material standards of living for a large part of the world's population. Nor do we encourage the Central Asian authorities to go back to the inward-oriented industrialisation policies of the past, based on distrust of global markets, as this strategy has failed to produce results elsewhere in the developing world.

Rather, this discussion is intended to encourage Central Asians to develop a balanced view on development and to equip them with arguments to steer national development policies in the right direction, and to start approaching the concept of development in a way that includes qualitative distinctions, still within the framework of free market economics.

Policy-makers should not think of development mainly in quantitative terms and in those vast abstractions – like GNP, industrial output, savings – which have their usefulness in developed countries but have little relevance to the development problems which appear in the countries of Central Asia. Instead, they should set specific parameters which take account of Central Asia's special conditions: its large natural resources, its dependence on land-locked pastures, its traditional societies who know the limitations of the land and climate much better than economic planners in distant capitals. On this basis they should devise appropriate development programmes which by nature will then benefit the people in rural areas – not upsetting their traditional ways of living but actually improving the quality of their lives.

Another purpose of this discussion is to expand the issue of environmental protection to include anyone who is involved in development – all relevant government departments, academic institutions, monetary authorities, NGOs, private industry and others. The issue of protecting the environment is of much too great a concern in this fragile region to be dealt with just by the Ministry for the Environment. Development programmes should be devised on an inter-ministerial and inter-disciplinary basis, incorporating both environmental concerns and economic and political realities. Only then can we remove the constant pressures on the environment and society exerted by short-term thinking and misguided development.

Of course Central Asians need to attract foreign investment, develop the export sector and enter international markets, because there is no way back. We cannot deny the Central Asians the opportunity to become full members of the global market, but this should take place in the wider context of what development is really needed – do they want everyone in the city to have Japanese-made Walkmans or do they want the rural population to be able to consume their own local products?

This dialogue should also involve international financial institutions and foreign donors. The current debate over sustainable development is based on the now widespread critique that many of the lending programmes and investments by major financial institutions have stimulated economic development by emphasising short-term cash-flow at the expense of longer-term, sustainable growth. Currently many of the multilateral agencies, such as the World Bank, are undergoing extensive internal reorientation towards a more socially and environmentally friendly profile. The national authorities in Central Asia should dare to debate with these institutions to achieve a definition of sustainable development which is appropriate to local conditions. In this process they can refer to some of the advances in economic theory made in recent years and point at the flaws of many of our current economic policies.

Ultimately, what is needed is a change of attitude. The economic principle of 'laissez-faire', as well as the ideology of communist state-planning, has wrongly created a mentality of taking things for granted, and we have become enslaved by the market and its monetary values. The alternative is not easy, in fact extremely difficult, because in effect it requires us to create a new economic model tailor-made to suit the conditions of our own societies. This can only be achieved if we are creative and feel responsible for the future of ourselves and our environment. This is what sustainable development is all about.

Notes

1 For an interesting discussion of the linkage between the theory of physics and economics, see *The Turning Point: Science, Society and the Rising Culture*, by Fritjof Capra, Bantam, New York, 1982.

2 See 'Changing Paradigms and Indicators: Implementing Equitable, Sustainable and Participatory Development' by Hazel Henderson, published in the book *Development: New Paradigms and Principles for the Twenty-First Century*, edited by Jo Marie Griesgaber and Bernhard Gunter, 1996.

3 Eastern Sichuan, inhabited by Tibetan people, used to be densely forested just a few decades ago. Economic pressures in over-populated Sichuan and eastern China have resulted in widespread deforestation. The author heard this saying on a trip through this region in September 1993.

4 See Note 2.

5 Environmental economists are working towards the improvement of economic theory, in the belief that many environmental problems can be solved within the quantitative framework of economics. While many environmental activists think that a clean environment should be obtained at any cost, economists hold that this has costs in terms of consumption of other goods foregone. By changing relative prices, qualitative indicators can be incorporated into the information on the basis of which we make

our economic choices. For example, by taxing products made by energy-consuming technologies we discourage the producer from continuing to produce in this way. Also, attempts have been made to quantify the value of, for instance, a national park, by estimating the amount of money and time people are willing to spend in visiting the park. Yet few economists believe that these estimates can provide the full picture. For how does one estimate the benefits of the national park on the overall environment of the planet?

6 An appropriate economic model for Central Asia should recognise the fundamental difference between the way traditional nomadic and semi-nomadic societies are organised and the most fundamental assumptions underlying economic theory. Central Asia's nomadic societies are community-based, in which (Buddhist, Islamic or Shamanist) values such as sharing and interconnectedness play an important role. In contrast, at the deepest level, classical economic theory is based on principles of selfish individualism: the more the individual consumes, the better off he is. Indeed, if GNP, as traditionally measured, is to grow, then each individual should consume more and more to boost demand and hence output. This leaves no room for the motivation of altruism, where the individual may incur costs for no conceivable benefit to himself. It thus reduces the meaning of community to a mere reciprocal arrangement among individuals: individual sacrifices on behalf of the community can only be seen as a personal investment or an insurance policy, for it will assure the individual that the community will help him in the future. Nor does it agree with the nomad's reluctance to accumulate consumer durables beyond a certain point, as he has to move around with all his goods. The traditional nomad even limits investment in mobile animals to the carrying capacity of the land. Likewise, the emphasis on consumption does not allow for religious-inspired abstention from consumption, which is based on the belief that the less attached one is to material things, the better off one is.

By simply assuming that the more one consumes the happier one is, classical economics completely overlooks the intricate relationship between consumption and the human mental experience. It is here, I believe, that economic theory meets with the other sciences of man, and it is here that we should look for the ultimate solution to man's problems in relation to his environment.

7 The author obtained this information on a field trip to Tongren county, Huangnan Prefecture in Qinghai Province in July 1994.

8 The countries of, for example, Mongolia, Bhutan, Kyrgyzstan, Tajikistan and Nepal make good examples, but it equally applies to Chinese provinces like Tibet, Qinghai and Xinjiang and inland Russian provinces.

9 To mitigate the problem of a small domestic market, Central Asian nations could consider the creation of a Central Asian common market. Several initiatives in that direction have already been taken: the Central Asian republics of the CIS have joined the Economic Cooperation Organisation (ECO), whose original members were Iran, Pakistan and Turkey. ECO has a natural identity as a bloc of all the non-Arab Islamic states of Western and Central Asia, but its functioning is hampered by Turkish-Iranian rivalry for regional influence. The provinces of north-west

China (Xinjiang, Qinghai, Gansu, Ningxia and Shaanxi) have linked up to assist each other in economic development.

10 This situation has been prevalent for the last few years, ever since Mongolia lost its material support from the former Soviet Union. Similarly, in Lhasa I heard of a shortage of wood, since most of the logs from eastern Tibet/western Sichuan were 'exported' to eastern China.

11 The author was shown around this future special high-tech zone on a fact-finding trip in June 1991. The local authorities referred to it proudly as 'Tibet's Shenzhen'. The same happened to him in the border city of Korgas, Xinjiang, along the Kazakhstan-China border, which has been designated as a 'Special Economic Zone' and was informally referred to as 'Xinjiang's Shenzhen', on a visit in June 1993. In 1991 some officials in land-locked Mongolia declared the whole country to be a 'special economic zone', similar to the ones on China's coast, but the fact that the isolated nation has so far failed to attract significant investors is a case in point.

12 See, for an interesting comparison between the economies of Central Asia and China, *The Economies of Central Asia*, by Richard Pomfret, 1995, pp. 136–7.

13 Even so, one should not underestimate the environmental damage caused as a 'by product' of these so-called successful development models; after the Asian financial crisis of 1997, which has exposed the down-side of the Tiger-model, such as severe social inequalities and pollution, one cannot easily maintain the truism that impressive economic growth necessarily and automatically improves the quality of life of the majority of the population.

14 The former-Soviet republics of Central Asia have reasonably good infrastructure as a result of the USSR's emphasis on electrification and education, but very poor transport networks for promoting export outside the CIS – all roads, railroads and air routes led to Moscow.

15 For a compelling description of the need for an appropriate development model based on appropriate technology and small-scale renewable energy resources, see *Ancient Futures: Learning from Ladakh*, by Helena Norberg-Hodge, 1992. The author witnessed the impact of modern development on this secluded Himalayan region over a period of twenty years, and draws far-reaching conclusions on the validity of our current economic policies.

16 The Aral Sea, which once was the world's fourth-largest inland body of water, has shrunk by almost a half and has now split into two. The land-locked sea was depleted by huge, ill-devised irrigation systems built by Moscow in order to irrigate cotton fields in Turkmenistan and Uzbekistan. Cotton was called 'white gold' for its ability to earn hard currency.

PART THREE

CASE STUDIES

Chapter 4

Sustainable Development

The Mongolian Experience

Z. Batjargal

Introduction

Humans need space, clean air, water, food, and other basic necessities to survive and maintain a good existence. But we live in a relatively closed life support system which, with the exception of energy from the sun, has finite resources. Although space and resources are limited, since 1950 we have added 1 billion people to an earth already confronted with unprecedented environmental problems. According to a UNFPA report, within the next half century we must anticipate a world population of nearly twice as many people, seeking three times as much food and clothing, and perhaps four times as much energy, while engaging in five to ten times as much economic activity.[1]

Today's world population is polarised between rich and poor. Much environmental degradation is caused by the world's wealthiest people, who harm the environment through their frighteningly high rates of resource consumption and pollution and waste generation. The world's poor also have a negative impact on the environment, by depleting their resources and generating waste and pollution out of necessity and the lack of alternatives for survival.

The effects of military activities must also be considered. According to the United Nations Environment Programme, global military expenditure reached more than US$ 1 trillion in 1990. Although military spending has recently decreased slightly as a share of gross national product (GNP), globally and in industrialised countries, it has increased in most developing countries. The military use vast amounts of mineral resources and energy, including 6% of the world's oil consumption (or nearly half the total oil consumption of all developing countries). Military conflicts also take their toll. For example, the 1991 conflict in Kuwait resulted in oil spills and extensive oil-well fires. The fires burned between 4 and 8 million barrels per day, an amount greater than Mongolia's total oil consumption over a period of two to three years. World political leaders, international communities, public movements and youth organisations should prevent the development of international political, social and ecological tensions before they result in life-threatening ecological disasters or armed conflict.

Humans, moreover, can alter their lifestyles and regulate private and social costs and actions to keep consumption and waste production within the limits of the natural world. In other words, sustainable development is possible. This implies both conservation and improvement of the natural world.

Population Control

Population control is often identified as an important step towards sustainable development, but it may not be the most appropriate approach because it is a largely intractable problem, and one which it is extremely difficult to enforce without impinging on basic human rights. On the other hand, an increasing human population requires more and more resources. This problem becomes even greater as increased affluence and lifestyle changes lead to demands for ever more luxury items. For example, if people in developed countries begin to want aeroplanes instead of just automobiles, people in the developing countries will also want aeroplanes.

Lifestyles

Therefore, in my opinion, the key problem is changing lifestyles. What do I mean by this? Since Mongolia has experience in changing lifestyles, I would like to offer some examples. In Mongolia, half of our people still live as they have for centuries; that is, as semi-nomadic herders. Their lives are closely tied to nature. Through the centuries, their lifestyles were largely in harmony with nature. They rarely exploited nature for more than basic, everyday necessities. This attitude resulted not only from a strong environmental ethic, but also from a lifestyle that provided for basic human needs without harming nature. However, over the last few decades, we began working to change traditional lifestyles and increase industrial development.

Mongolia wanted to become a more developed nation in a very short period of time. While this resulted in many positive changes, particularly in education, public health and culture, in terms of economic growth and several other indicators of quality of life, we cannot say the same, especially if we include social costs. For example, at present, to produce grain products we require very high investments for very low yields, compared with traditional agricultural practices (see Figure 1).

As Figure 1 illustrates, traditional practices are far more cost-effective and ecologically sound. In addition, traditional practices generally provide cleaner, purer products. Modern agricultural practices require more chemicals and produce lower-quality products. Furthermore, large areas cultivated as a monoculture are more vulnerable to pests.

Another example is the production of livestock products. From the 1930s to the 1950s, Mongolia began collectivisation of its livestock industry, and by the 1960s, most livestock was the property of the collective. At the same time, we introduced industrial methods of livestock production. The main goal was to maximise products, while minimising investment. We therefore used methods which concentrated and specialised livestock production and centralised management.

Figure 1: Comparison of Modern Technology and Traditional Practice in Agriculture

Modern Technology	Traditional Practice
Cropland	Cropland, used as pasture during part
Irrigation construction (e.g. dams, canals)	of the year (saves pasture and
Irrigation machinery	provides organic fertiliser)
Heavy agricultural machinery	Horse or camel (no machinery, no
Gasoline	gasoline, no engineers, etc.)
Mineral fertilisers	Simple, hand-made, yet effective,
Chemicals (e.g. pesticides)	plough that causes little erosion.
Energy for machinery	
Spare parts for equipment	
Special staff (e.g. engineers)	

Outcomes

Crops	Crops
Soil Erosion	Livestock Products
Soil Pollution	No Erosion
Water Pollution	No Pollution

95

What were the results? In some areas livestock concentration resulted in numbers exceeding the carrying capacity of the pastures, resulting in overgrazing. Specialisation caused ineffective use of pastures and overgrazing of preferred forage. At the same time, specialisation did not serve the needs of the local people. Traditionally, Mongolian families owned several different species of livestock which they used for different purposes. For example, cows were raised for milk, sheep for mutton, horses for traditional drinks and to herd other livestock, camels for transportation, and goats for children's dairy products. This was not only more convenient for local people, but the different species of livestock formed a community that utilised pastures more efficiently. Rather than compete, these species utilised different forage plants, thereby complementing one another. The construction of shelters for livestock helped them survive during extreme weather, but animals kept in shelters gradually lost their tolerance for inclement weather and the quality of their wool decreased. Industrialised agricultural methods which worked well with highly productive breeds of livestock imported from abroad did not produce the same results with locally adapted breeds of livestock. As a result of our investment, the traditional, self-sustaining and relatively stable ecosystem, of which our livestock were a part, was transformed into a more fragile, successively younger ecosystem. Such systems require more external support for their maintenance.

Let me provide a third example of the problems we encountered when we tried to modernise industries. Traditionally, every family in Mongolia functioned as a small manufacturer. To make the products they required, Mongols used materials provided by nature, such as wood and water. After using these products, they returned the materials back to nature in forms easily degradable and incorporated into natural biogeochemical cycles. This was an almost waste-free technology that rarely disrupted natural cycles significantly. In contrast, modern technologies and industries require enormous quantities of raw materials and result in wastes that cannot be easily returned to nature. Instead, these industries require landfills or incineration. And even landfills and incineration are inadequate for the disposal of hazardous and toxic wastes, especially radioactive materials. This means we lose twice with modern technology: first we disrupt the environment to acquire raw materials, and then we are faced with the problems of waste disposal. Alternatively, in the case of traditional practices, losses were close to zero.

Mongolian traditional lifestyles were replaced by so-called socialist lifestyles and associated public property rights. Today Mongolia is again in a transition period, as it attempts to move away from a centrally controlled society towards private property rights, a free market economy, and political and ideological pluralism. But again, the new social system is based on an industrialised societal tradition, this time a social system dominated by individualism. We therefore face new problems. Already there are indications that individuals are attempting to maximise their returns from common pool resources, such as ground water, pasture and air. For example, individuals are grazing additional livestock on common pastures. The same situation occurs with ground water use and air pollution. Individuals have little incentive to protect the land or manage resources for the long-term, beneficial use of all.

How were such problems managed in the past? First, people did not require more production beyond that necessary for their families. Second, people had a strong environmental ethic, with associated social mores so strong and pervasive that they functioned as informal, customary laws. This customary use of common pool resources reflected the interests of the entire community. Mongolian tradition never allowed individuals to believe they were separate from the community. As children, Mongols learned that individuals were part of a larger whole; therefore, benefits which accrued to the entire community also accrued to the individual.

The other problem we face today concerns the mechanism of market economies. A fundamental criticism of modern, free market economies is their inability to consider all the impacts of transactions; that is, the inability of free markets to adequately internalise externalities. For example, the interests of non-humans, such as plants, animals and ecosystems, and indirect human participants, including future generations, are not reflected in market transactions.

Environmental costs are not internalised; therefore, the price of a product almost never reflects its true cost to society. In Mongolia we realise that it will be difficult to internalise market externalities because even environmentally aware consumers in highly developed market economies find it difficult, if not impossible, to make environmentally responsible decisions.[2] Many people believe that modern science and technology will help us overcome these obstacles. But the science-based civilisation that has dominated in the 20th century, and a false faith in science and technology, have resulted in dramatic environmental changes. For many, faith in science has replaced other values, including religion and tradition. Historically, religion played a crucial role in defining the relationship between humans and their environment. In practice, however, none of the world's major religions has successfully motivated its adherents to exercise sound environmental practices.

Changing Human Consciousness

At this point it is legitimate to ask 'What are some possible solutions?' One option, especially in Mongolia, is to go 'Back to the Future'. This does not mean we should strive to return to conditions which existed under socialism or the empire of Genghiz Khan. Rather, it means that we should learn the principles of our traditional ways of thinking about nature.[3] History shows us the roots of many present-day problems in several areas of the world. I am not arguing that rejecting modern technology will prevent environmental degradation, only that traditional ethics toward nature may help us find ways to use this technology in a manner that minimises its destructive impact on nature and the environment.

Given this background, what are some concrete proposals for addressing sustainable development and environmental conservation? First, we should recognise that the goal of maximising production often conflicts with nature conservation, as such a goal rarely includes consideration of the natural world. Second, since people are often positioned at the end of a food chain, humans will often suffer the most from toxic substances introduced into the environment. A more non-anthropocentric approach to the environment should be developed, since nature, including humans, is itself a source of values. Humankind should make a concerted effort to change patterns of consciousness toward more environmentally friendly ethics, values and attitudes. For example, old Mongolian traditions include a naturalistic ethic which respects all forms of life and includes all life forms as a part of a larger, sacred whole. A similar awareness must eventually pervade all human consciousness and become part of a more universal ethic, before new, more sustainable economic theories are developed which offer a significant challenge to the dominant economic paradigm.[4]

Changing Consumption Patterns

Achieving sustainable development will require changes in consumption patterns. In many instances, this will necessitate reorientation away from the production and consumption patterns which developed in industrial societies and have been emulated throughout much of the rest of the world.[5] To address mounting quantities of waste, societies must facilitate recycling, reduce wasteful packaging, and encourage the introduction of more environmentally sound products and practices. Traditional practices can often show the way. For example, traditional Mongolian practices employed only reusable and completely biodegradable packaging. In contrast, industrial societies needlessly create additional waste. Moreover, products in industrial societies are often packaged in materials which cost more than the product itself.

Regulation

Conflict often arises over government policies and programmes, especially legislation and regulations that affect business. The technique used to make a decision can often strongly influence whether or not the decision achieves its objective and is accepted by those who have a stake in the outcome. During the development of regulations, decision-makers should attempt to consider real-life situations and avoid sources of conflict with local inhabitants; otherwise they risk passing ineffective or unenforceable regulations. A good approach is to learn local values, attitudes and customs and apply these principles to more formal regulations. For example, in Mongolia issues like land, soil and water protection were until recently solved in relatively simple ways. Local customs and regulations were not based on abstract religious warnings, but on practical, everyday threats. For example, children were taught not to dig needlessly in the ground, because to traditional Mongolians this was the same as cutting into their skin. As a result, in the past there was no need for large regulatory structures and organisations, because actions were controlled by powerful and pervasive social ethics and mores.

Science

Traditional practices and the knowledge and wisdom accumulated by local people have been tested over many centuries and therefore represent a great source of information for science and technology. In fact, traditional practices and knowledge are, in essence, the outcomes of long, iterative experiments. During these experiments, humankind lost civilisations in all parts of the world as people competed and conflicted with nature. The lesson is relatively straightforward: instead of using science and technology as tools for mastering nature, we should use both to help us to understand natural processes better and to find ways of harmonising our activities with the natural world.

Resource Use and Solid Waste

If we obtain comprehensive knowledge about natural cycles and processes, we should be able to develop technologies which simulate natural processes. For example, modern recycling technologies should mimic natural recycling processes. Only in this way will we be able to transform current technologies which disrupt natural cycles several times, into more sustainable technologies which cause little or no disruption to natural cycles.

Diversity

We need to maintain and even increase diversity; not just biological diversity, but also diverse lifestyles. We must avoid homogenising landscapes and taking monoculture approaches to agriculture. Because the natural world is so biologically diverse, it is important to maintain a diversity of lifestyles when searching for optimal approaches to sustainable development. However, we must be able to integrate these diverse lifestyles into sustainable development programmes for the common benefit of nature and human societies.

International Environmental Standards

It is necessary to change the definition of international standards on the quality of drinking water, milk, food, air, etc. For Mongolians, pure or distilled water is not the most desirable drinking water. Water treated, for example, with chlorine, is even less desirable. Instead, natural spring water containing some minerals, salts and even microorganisms is the highest-quality water. If international standards, developed for industrialised countries, were enforced on all nations, people in many parts of the world might well have to rely on artificial, rather than natural, sources of drinking water. Similarly, high fat content is viewed as unhealthy in several industrialised nations, such as the United States, but in Mongolia, the higher the fat content, the higher the quality of the meat. When developing international standards it is important to consider the lifestyles, customs and traditions of people from all nations, to ensure that imposed standards do not result in the natural world becoming more homogeneous and uniform.

Carrying Capacity

Although we usually use the term 'carrying capacity' solely with respect to human and ecological populations and communities, we need to expand its meaning to include natural cycles. For example, when we discuss carrying capacity with respect to a particular parcel of land, such as a pasture, we should expand our discussions beyond the numbers of livestock it can carry to include the capacity of the pasture to maintain its natural cycles. This implies expanding our search for new approaches to sustainable development, to include approaches which address problems associated with resource extraction and waste disposal simultaneously. In other words, we should search for process-oriented approaches to sustainable development which consider the capacity of the environment's most important natural cycles.

Protected Areas

The International Union for the Conservation of Nature's concept of protected areas focuses primarily on the scientific value of protected areas. Specifically, the IUCN considers protected areas as places to conduct research, monitoring, observation, etc. But in my opinion, protected areas should be established primarily to support the maintenance of natural equilibria and biospheric capacities and to provide refuges for rare and endangered species of flora and fauna. In Mongolia, when we select areas for the conservation of nature and natural processes, we attempt at the same time to protect areas which are sacred to local inhabitants. This facilitates conservation of the protected areas, because traditional criteria for sacred areas were based on natural features and characteristics. Special organisational structures for protecting such areas therefore become largely unnecessary. Today's societies could learn much from such approaches.[6]

Conclusion

In conclusion, I would like to stress that traditional practices and lifestyles may not provide instantaneous solutions to the problems currently facing modern society. They can, however, provide the foundation for new approaches to conservation and sustainable development. To achieve sustainable development, nations can use new approaches to solve old problems or old, traditional approaches to solve new problems. However, a combination of both old and new approaches to sustainable development would provide the most innovative solutions to current environmental problems.

Characteristics of Central Asia today

- High annual and diurnal temperature fluctuations
- Low rainfall
- High evapotranspiration
- Fragile natural ecosystems
- Widespread soil erosion and land degradation
- Low rates of soil humus production and vegetative regeneration
- Low carrying capacity of pasture land
- High grazing pressure
- Intensive deforestation
- Limited water resources
- Low quality of drinking water (high level of mineralisation)
- Ecologically unique areas and ecosystems
- Rich wildlife resources including rare and endangered species
- High population growth
- Rapid economic changes and development

What was Central Asia in the past?

- Nomadic civilisation
- Vigorous and well-adapted people
- Lifestyle harmonised with nature

What could Central Asia be in the future?

- The model area of sustainable development based on traditional lifestyles and modern technologies.

Notes

1 UNFPA, *Population, Resources and the Environment: The Critical Challenges*, London, 1991.

2 Zachery A. Smith, *The Environmental Policy Paradox*, Prentice Hall, Englewood Cliffs, New Jersey, 1992, p. 12.

3 Z. Batjargal, 'New Challenges, New Solutions', presentation to the Fourth Annual Pacific Environmental Conference, East-West Center, Honolulu, Hawaii, 27–29 March 1994, p. 8.

4 Rogene A. Buchholtz, *Principles of Environmental Management: the Greening of Business*, Prentice Hall, Englewood Cliffs, New Jersey, 1993, p. 60.

5 M. Keating, *The Earth Summit's Agenda for Change*, Centre for Our Common Future, Geneva, 1993, p. 7.

6 Z. Batjargal, 'Sustainable Development with Innovative Environmental Protection and Management', presentation to the Environmental Conference in Potsdam, Germany, 3–5 June 1994, unpublished.

Chapter 5

Amalgamating the Free Market and Traditional Nomadic Society
Sustainable Economic Development for Mongolia

Alicia J. Campi

In the summer of 1992, at a Mongolian Studies conference in Ulaanbaatar, I presented a paper arguing that the standard economic development models being designed for Mongolia by Western experts would meet with only limited success, because they did not consider the country's environmentally-attuned, nomadic society and value system as key factors. Reactions to my speech among foreign experts ranged from curiosity to scepticism, since it challenged the universal applicability of free market theory to diverse countries and peoples. However, my paper appeared to strike an emotional chord with Mongol colleagues, whose responses were overwhelmingly enthusiastic and positive. My speech was subsequently translated into Mongolian and published in a local newspaper and a research journal.

In 1993 I expanded on my original premise for a chapter entitled 'The Special Cultural and Sociological Challenges Involved in Modernising Mongolia's Nomadic Socialist Economy', in a book honoring the American Mongolist Dr Henry Schwarz. This was a scholarly attempt to illustrate the necessity for economic planners and policymakers, foreign and Mongolian, engaged in reforming and modernising Mongolia in the post-Communist era to research the fundamentals of the historical Mongolian/Central Asian model of nomadic social and economic relations.

During the past three years my experience of operating a joint-venture consultancy specialising in Mongolia has also greatly affected my views. It is one thing to theorise about sustainable development in Mongolia, and by extension in other Central Asian nations (as I did when an academic and later a US diplomat), but altogether a different matter to run a business and act as an intermediary promoting commercial and educational exchanges and joint ventures between Mongolia and the United States. It is because I have experienced the impact of Mongolia's nomadic heritage on present-day business relationships in Mongolia's public and private sectors that I have reached the conclusion that classical economic theories of development have only partial applicability to Mongolia. Thus it is valid for forums such as this conference seriously to examine whether an improved definition of sustainable economic development is warranted for Mongolia and other traditional Central Asian economies.

Mongolia is a landlocked country of 2.4 million people and 25 million head of livestock. It is still economically dependent on the style of animal herding that was developed by the peoples scattered over the hostile steppe and desert lands of Central Asia. This special nomadic economy, intimately tied to the beautiful yet harsh environment, did not die out even in the modern era with the development of urban centres. During the socialist period in the 20th century, the Mongol government, spurred by its mentor, the Soviet Union, emphasised urbanisation, light industrialisation and agriculture at the expense of the traditional animal husbandry base. Educated Mongols in the cities attempted to cast off the mentality of their nomadic pre-revolutionary social system in favour of scientific Communism. Nevertheless, seventy years of Soviet-inspired social and economic policies did not destroy the uniqueness of certain national social characteristics, referred to as a 'peculiar inertness or stagnation' by Russian Mongolists such as L. N. Gumilev.

In 1994 Mongolia's command economy is well along the path of transformation to a free-market, democratic system. Western economic advisers are pressuring Mongolian government officials and businessmen to adopt the new 'foreign economic religion' of free marketism. The nomadic nature of Mongolia's traditional economy is dismissed, if it is even considered. Instead, economists point to the successful development of Asia's 'Four Little Tigers' or lump Mongolia's situation with that of the Eastern European states.

Such specialists fail to recognise that modern free-market theories, not unlike the discredited communist ones, arose from the sedentary, agricultural and industrial experience of Western countries. The apparent universality of these principles among countries in other areas of the globe which have developed successfully during this century is due to the fact that these economies also rest on a sedentary, agricultural base. Moreover, many of the Western economic experts who have been commissioned to provide guidance for Mongolia's transition to a market economy are well versed in Soviet or Chinese-style command economies, which are also rooted in these same economic motifs. The conditions of nomadic life are unreal, even unimaginable to them. After all, it is very difficult for people who live in industrialised, urban settings to understand the dynamics of nomadism. We simply do not believe that some people would choose to live in tents, always moving around with their herds. Consciously or subconsciously, most of us hold the view that the nomad is a barbarian – a perspective known as the 'barbarian syndrome'. In the West as well as in most developing countries in the world, indigenous nomadic populations were killed off or forcibly assimilated over the centuries. Today in Europe the remnant nomadic group known as Gypsies are reviled and unabashedly discriminated against, as seen in the 1992 expulsion of 60,000 Gypsies back to Romania by the German government. In the United States native nomadic Indian populations have been isolated for decades on special reservations, severely limiting tribal migrations.

Such hostility towards nomads is especially intriguing when one considers that the emergence of pastoral nomadism across the Eurasian steppes is one of the most frequently noted examples of ecological adaptation in human history. Pastoral nomadism has dominated the Inner Asian steppe for centuries as a sophisticated economic specialisation based on horse-riding, which has successfully exploited the land's limited resources. Nonetheless, Mongolian nomadism has always engendered distrust and fear in the country's two great neighbours. The Chinese traditionally viewed the horse-riding northern border peoples of Central Asia and the Mongolian plateau as fundamentally different from and hostile to them. Chinese history is perceived as a struggle for dominance between the sedentary, agricultural, that is to say 'cultured', Chinese, and mobile nomadic tribes such as the Mongols. The typical Russian aversion to nomadic peoples is evident in the colonisation of the steppe lands of Central Asia. Tsarist and Stalinist policies over several centuries aimed at turning nomadic confederations against each other, then settling the nomads down, and finlly intermarrying them with Russians. The legacy of such policies is evident in the hybrid societies of Kazakhstan, Kyrgyzstan, and Uzbekistan.

Mongolia, as the ultimate buffer state between its giant neighbours, was consciously spared large-scale colonisation and modernisation by China and Russia. At the time of its Communist revolution in 1921 Mongolia had no working class in the modern sense. In fact, Marshal Choibalsan observed that in the mid-1920s there were not more than 150 Mongols engaged in any kind of industrial activity. Few Mongols were involved in trades or crafts that could be adapted to modern industrial life, because the vast majority were livestock-herding nomads or Buddhist lamas. With no surplus urban manpower to utilise for industrialisation, tens of thousands of Soviet and Eastern European advisers were brought into Mongolia for decades to operate the few major state enterprises, factories and mines. Even by 1990, when the socialist system began to disintegrate, Mongolia had not created sufficient numbers of skilled native industrial workers, because of the preference of educated Mongols for jobs in the Government bureaucracy in the capital, and the reluctance of countryside Mongols to leave the livestock economy.

Marxists of all ilks have been troubled by nomadism: pastoral nomads do not fit well into any unilinear historical stages. Throughout history, when nomadic states collapsed the Mongolian nomads appeared to revert to their traditional tribal organisation, which should be impossible in Marxist theory, if the tribal institutions really were destroyed during state-formation. The Marxists' open disregard for the economic utility of Mongolia's nomadic economy is continued today by Western experts; both groups believe that integration of the pastoral economy, with its thin dispersion of people, and the creation of a produce-intensive agricultural economy, is inevitable. Even renowned Mongolists such as Owen Lattimore believed that industrialism could bridge the chasm between agricultural society and Mongolia's pastoralism.

Decades of foreign negativism has convinced many of the Mongol intelligentsia that there is no future for nomadism in Mongolia. Urbanised Mongols are alienated from the nomadic lifestyle. While mouthing respect for traditional nomadic life, urban dwellers until recently have not encouraged their children to take up the life of an animal herder because they accepted the Socialist view that the herdsman was conservative and backward, while the working proletariat and bureaucracy were necessarily more progressive. The recent phenomenon of young people migrating to the countryside has been caused by collapsing economic opportunities for them since the dismantling of the command economy. Moreover, with the rise in urban crime, both the government and citizenry believe that social order will improve if young unemployed people return to their countryside relatives who are busy with their private herds rather than remain in the city carousing and getting into trouble.

Under Mongolia's Communist system three classes developed: the *arat* herdsman, urban workers, and bureaucratic intelligentsia. The third group formed the political apparatus, a new elite with higher wages and real power, who replaced the pre-revolutionary nobility and religious hierarchy. Their high economic and social status made them the chief beneficiaries of the socialist political system, and gave them a vested interest in the maintenance of the regime. Gradually they formed a new state administrative structure, increasingly divorced from the realities of the animal husbandry economy, but still economically dependent upon it. It is important to recognise that today's Mongol political leaders and businessmen are being drawn mainly from this class. These are the individuals foreign experts meet for discussions and expose to free market training programmes. Yet despite their modern façade and measurable alienation from their compatriots in the countryside, how Westernised are they in reality? And how much are they still operating with values and principles handed down from their Mongolian nomadic tradition?

The nomadic economy's sociological and cultural value characteristics do persist in modern Mongols of the town as well as the countryside. The history of Mongolia's struggle against its neighbours has left a feeling of ethnocentricism in their souls, which makes change in the modernisation process difficult to achieve. There are many examples of this clash in values; I will highlight only a few.

Let us begin with the concept of a work ethic. Mongols are commonly criticised by foreigners, whether Chinese, Russians or Westerners, as being lazy. One need only refer to an interview with a German economist in the *Mongol Messenger*, an English language newspaper in the capital in July this year. Dr Dieter Benecke commented: 'What I mean by this is a certain lazy way of not really being interested in serving people and expecting something will happen rather than taking initiatives and making efforts.'[1] He guesses this is a legacy of the Soviet period, but such comments are pervasive in prerevolutionary accounts by foreign visitors to Mongolia. It is true that Mongols are not clock-watchers and punctuality is not a virtue in this society, but no one works harder under severe conditions than the herdsman to save his animals during storm or at lambing time. No one is more constantly alert to changing weather or wolf attack. Nomadic life calls for periods of intense activity followed by much 'down time' watching animals. Such a working style is similar to that of a farmer, and farmers are not accused of being lazy.

One must remember that as late as 1968 two thirds of Mongolia's population were still engaged in animal husbandry. With this cultural attitude towards time, Mongol workers were then exposed to Soviet-style management concepts that paid little heed to productivity and cost reduction. Mongol dislike of regimen and regularity is falsely interpreted by foreigners as laziness. During the Socialist period light industrial enterprises used negative incentives including labour camps and pay reduction to enforce discipline. In Ulaanbaatar this year there has been a serious controversy over restricting toilet time and other working conditions in a British-Mongolian sewing joint venture. This problem is a reflection of different work ethics and the managers' loss of the right to use the stringent punishment system of the Communist era to enforce factory regulations. Finding an acceptable way of modifying the traditional Mongolian work style to meet industrial needs is a problem still to be solved, but should be approached by foreign joint venture partners with flexibility, using positive as well as negative incentives that are culturally compatible.

Another major clash in cultural values between Mongols and foreigners is the attitude towards land-holding. For Mongols personal mobility and the portability of shelter and household goods are the key needs for survival. Theirs is a repetitive, cyclical, migratory existence in the endless search for pasture. Because the climate is not conducive to agriculture, supplementary fodder is rare. The nomad exploits the grazing land in a highly organised, usually environmentally sensitive way, with precise division of skilled labour, definite assignment of responsibilities, and utilisation of the extended family in cooperative economic activities. Traditionally, there has been little competition for land *per se*, but great controversy over land usage. For nomads, time and space are intertwined aspects of ownership. Exclusive land ownership, which is such a strong economic motivator in agricultural societies, has no intrinsic value since no pasture can be grazed continuously. This lack of strong interest in owning land constitutes a vast difference from nearly every sedentary society. One of the mainstays of the free market system is the freedom to buy and sell immovable property. Western economists, schooled in landfast societies, never question the premise that all societies share the desire for property ownership, which then acts as a strong motivator to save money. When we look at Mongolia's three-year history of privatisation, we see that state enterprises were easily sold, by comparison with ex-socialist states in Europe, and a stock market was quickly established. Yet private home- and land-ownership has sill not been implemented – pasture land in fact is explicitly exempt from purchase by constitu-tional law, and there has been no great pressure from Mongolian opposition parties to change present land issue policies quickly.

Another area of clashing socio-economic values is the field of commercial relations. Domestic trade within Mongolia has always been limited in scope. Nomads traditionally used barter exchange among themselves if necessary. Animal products from their herds were also traded for the commodities of sedentary people, such as tea and salt, by wandering pedlars; but raiding settlements was just as important a source of necessities for the nomads. It is this pattern of pillaging that caused settled peoples to fear and loathe nomads the world over. Trade was always in the hands of tribal leaders, who engaged foreigners to conduct mercantile activity. Historically, the Mongols had commercial relations with the Chinese agricultural peoples south of the Great Wall, Central Asians further west, Persians, Russians and Arabs; the volume of trade ebbed and flowed with nomadic political power.

It is crucial to emphasise that unlike other societies, no matter how rudimentary, Mongol nomadic society did not naturally develop an indigenous merchant class, nor were there any trade markets until permanent religious communities arose in the 16th century. Rather, trade was conducted by notorious foreign itinerant traders, known for their exploitative, usurious character. This system persisted throughout Mongol history, from the Empire period, through the Manchu-Chinese epoch, to the recent socialist era when commerce was dominated by Soviet and east-European monopolies. The Manchu pattern of letting foreign traders go out to nomadic encampments with their goods instead of encouraging nomads to trade at open border markets was perpetuated at Soviet insistence by the Mongol government through its own state monopoly procurement system. Large Mongol state trade organisations such as Mongolimpex acted more as brokers than traders. Commercial decisions such as the prices of minerals were often made under pressure from the Mongol Politburo or direct interference from Moscow.

Thus trade has come over the centuries to be viewed by Mongols not as a valid way to build wealth, but as the instrument of exploitation. During the Manchu period, Chinese traders lent Mongol princes money at high interest rates which were only passed on in the form of onerous levies on their subjects. With the great influx of foreign capital and foods, Mongols of all classes became seriously indebted to these traders by mortgaging future livestock production for immediate consumption of luxuries. Despite being dependent on trade for tea, tobacco, silk and grains, Mongols despised the Chinese merchants who provided them as liars and cheats. This history of exploitation generated strong negative attitudes towards commerce in general, and an ever-present fear that trade with China would lead to political dominance and assimilation.

With the collapse of the Soviet Union and the CMEA (Council for Mutual Economic Assistance) trade system, Mongols have analysed the past decades of Soviet economic assistance, which led to the present large foreign debt. Feelings of having been exploited by foreign traders have again arisen among the Mongols. There is no doubt that the inflated or undervalued barter trade terms which underlay the CMEA system caused the Mongols to have a poor understanding of the true market value of their export products. This great naiveté, as well as greed, is behind the major mistakes made in the international money market scandal called the 'Gold and Dealers Case'. Even today the authorities are hesitant to move from barter purchase to cash payment for Mongolian minerals. Many of the major foreign transactions are still tied to assistance packages (such as the Japanese one), which are based on the concept of barter.

Because barter was the main form of exchange in the nomadic economy, Mongolian use of money, and ideas about its importance, are significantly different from those of the Western world. The Mongols did not develop their own national currency, the tugrik, until 1925. Paper currency was distrusted; the preference was for silver or commodities. The first bank was established in 1924: goods were bartered directly, or commodities such as sheep, tea bricks and salt could be exchanged for foreign silver dollars. Reformers attempting today to modernise Mongolia's banking system, with its 20%-30% monthly interest rates, must recognise that the average Mongol's negative attitude towards banks and the entire lending process will not be easily overcome.

The Communist government established a monetary economy in Mongolia not for the nomadic majority but to benefit the small urban elite. Currency only serves very limited functions in a nomadic economy. Wealth is accumulated in the form of living livestock, exchanged only when goods are needed. Migratory life necessarily means carrying few material possessions, so Mongols are not very oriented towards consumer goods. Money is only another possession with little value, since the herdsman can feed, clothe and house himself from his herds. The resulting mentality is that money is not to be saved but spent quickly, even frivolously. Another factor degrading the importance of money in Mongolia is that the salary range of sedentary workers during the Communist era was not great. The most highly-paid Mongols received only three times the pay of the least skilled workers. Position and power were reflected in other benefits. The pay differential was less than in any non-communist country in Asia. In a recent article in the *Mongol Messenger*, the nomadic concept of monetarism is mentioned and the view is presented that the lack of money in the hands of both producers and consumers will mean continued economic decline instead of revival.[2]

In today's economy the herdsman, who represents over 30% of the population, is still a major holder of economic power because he is the source of food, and the second largest supplier, through his animals' hair, wool, meat and by-products, of hard currency for the nation. The ranks of the livestock producers have been growing, not decreasing, since privatisation and economic transition. Yet there is little incentive in the structure of the Mongolian economy to encourage him to turn his animal wealth into cash. In the countryside money can be used to purchase rice, tea, clothes, alcohol and sweets, but bartering still serves the same purpose. The migratory form of subsistence is oriented towards immediate consumption. Goods are acquired and consumed as needed, with no incentive to accumulate food, material possessions or money. On the other hand, animal wealth is mobile and a convenient form of savings.

The assumption is widely held by economists, though rarely articulated, that all human beings wish to accumulate money, and that currency is a prized entity. Such an assumption arises from the sedentary economy experience, where money serves a useful purpose. Since the value of currency is quite limited in a nomadic economy, it is not the motivating economic stimulus modern economists subconsciously believe. That is why many free market price reform models and supply-and-demand theorems may not work in Mongolia.

For example, at a seminar in the US in 1992, devoted to assisting Mongolia's transition to a market economy, the Mongol participants were lectured to by an American economist on the benefits of floating milk prices. The Mongol official who had implemented just this policy under pressure from Western economic advisers the year before was listening in the room, and complained that after letting milk prices in Ulaanbaatar float for six months, the price of milk in the city had increased nine times but the supply had been reduced by a half. These milk shortages had created widespread urban unrest and great political pressure was exerted on the government. The American professor's response was that either the policy had been wrongly implemented (an example of the 'Mongols did it wrong' syndrome), or the policy had failed to take into account the desire of the herdsmen milk producers for increased leisure after earning high profits for their milk. The inappropriateness of such explanations lies in the fact that the animals must be milked daily, regardless of the milk price. The herdsmen were certainly not abandoning their daily herding activities to enjoy themselves. The milk price control board had not mismanaged the price reforms at all; foreign advisers and urban Mongol officials alike had failed to realise that the milk producers had little use for the money they were receiving. Since the government had placed only a few consumer goods in the countryside for the nomads to buy, and furthermore the nomadic lifestyle itself precludes the consumption of material goods for its own sake, there was not much incentive for the herdsmen to save their increased profits. They had no need to save money for their children's higher education or to purchase land or a house – all motivators of no meaning to a nomad. This is a classic example of people with earning power having little reason to sell their product to the urban consumer because the form of payment is unattractive to the producer. But where was all the milk going? It stayed in the countryside, where nomadic families increased their dairy consumption and turned the excess into dried milk products for use later in the year. This failure of free-market price reform was directly related to Western policy advisers' ignorance of how the nomadic economy functions in the cultural context of Mongolia.

As mentioned above, this lack of understanding extends to urbanised Mongols as well as foreign economists. Just this year Mongolians and foreign traders have sought to use new Soviet jeeps, going for around US$7,800 in Ulaanbaatar, as a medium of exchange for around 600 sheep in the countryside. Takers for this offer have been few – why? First of all, the jeep is only as good as the petrol supply. In the countryside, gasoline is expensive, at over US$1 for 2 litres, and in very short, irregular supply. If there is no gas in the city, the urban dweller can lock up his jeep and wait days or weeks for more fuel. But what is the migrating herdsman to do? His animals must move constantly to find new grass to survive. Should he abandon the gasless jeep or just pull it along by camel and ox? Secondly, this transaction places the value of a sheep (at about US$13 a head) far in excess of its value on the world market. At the end of August this year in the Chicago market, spring lamb was priced at US$0.72 a pound. That is in an economy far more expensive than Mongolia's. In the just-released Mongolian State Statistical Office figures, as of May

1994 the monthly family income in the countryside was 32,900 tugriks, or about US$80, which equals US$960 yearly. This means that at US$13 a head, one sheep would represent 1/6 of an average herding family's monthly income, which reflects a very skewed notion of the value of sheep in the Mongolian economy. These examples illustrate the sociological and economic factors impacting on Mongolia's economy today. Foreign experts' flawed understanding compounds the situation. A typical error is to see the source of all Mongolian economic problems in the imposition of a command economy during the socialist era. Such a view presumes that a rudimentary market economy existed in Mongolia, as it did in Russia, Czechoslovakia, Hungary and other states prior to the introduction of communism. The historical record proves that such a presumption is false.

Another incorrect premise is that Mongolia's nomadic economy is comparable to ranching in developed countries. According to this view, Mongols are, or should be, the same as ranchers. Even well-known Mongolists have made this error: Owen Lattimore proclaimed in 1940 that mobility was not a necessity in Mongolia, and that the economy could be pastoral but not nomadic. Yet the geography and climate of Mongolia preclude a full-scale move to ranching or sedentary pastoralism. Without expensive outside support in the form of imported fodder and shelter for animals, the harsh climate and sparse vegetation forbid Mongols from living like dairy farmers. Nomadism could be abandoned if the herdsmen changed from grazing their animals to fodder feeding. Since the country cannot support much normal agriculture, where is the fodder to come from? And where are the financial resources to pay for imported feed?

Beyond these economic factors is the foreign sedentary world's failure to understand that Mongols still cherish and voluntarily perpetuate nomadism, an economic form that requires long periods of isolation, large investment in herds and inventory, marginal material wealth, and constant exposure to harsh weather conditions. The bottom line is that we non-Mongols find it nearly impossible to accept that the herdsmen want to retain their nomadic life and reject settled, urbanised, industrialised society. The American scholar George Murphy has written that Mongols possess a 'high willingness to live in a society of stable poverty'. The Mongols' genuine attachment to their traditional herding techniques and nomadic lifestyle emerges as a repetitive theme in the writings of explorers and scholars of Mongolia now and in the past, but modern economists still dismiss and ignore this cultural value. We need only recall that in the past the Mongols could see the supposed economic advantages of the Russian and Chinese sedentary economies, yet still chose to continue nomadising. In the regions where Mongols became sedentary, this was forced upon them by Chinese and Russian policies. In the 20th century low-income Mongols did not move to the cities to adopt a more comfortable sedentary life. The migration was spurred by the politically oriented education system, which was centralised in settlements, reviving the role of monasteries in the pre-revolutionary period. As adults these educated Mongols could only find work in the communist bureau-cracy or urban state-run enterprises, and with time they developed a taste for sedentary living. The less educated, less politically-connected nomads chose to remain in the traditional economy rather than take unskilled jobs in the cities. This eco-cultural pattern is fundamentally different from economic development in agricultural societies.

Conclusion

Where should Mongolia look for realistic economic strategies for sustainable development as the country moves into the 21st century? One possible strategy is for Mongolia to study its other Central Asian counterparts, particularly those nations with the same nomadic and socialist experiences, to exchange views on similar problems, strengths, cultural values and conditions. Central Asia as an economic unit is strong in mineral resources, but its landlocked position and relatively low population density due to its climate and environment preclude easy economic development solutions. Mongolia can work with its Central Asian neighbors to create a new strategy for sustainable development based on preserving long-held cultural and economic forms.

Mongolia should also look to the North East Asia Free Trade Zone concept as a way to ensure an outlet for its products to the sea, and as an entry into the fast-growing Pacific Basin market. Other Central Asian nations may see their best options in trading alignment with the Middle East and Europe, but it is clear that Mongolia's economic destiny should be linked to the Asia-Pacific region. While it is appropriate for Mongolia to investigate its potential role in a North East Asia Free Trade Zone, which would tie it to the nations of the Pacific Basin, it should not forget to look westward as well, because the other Central Asian nations which have emerged from the break-up of the Soviet Union may have more in common with Mongolia's development strategy than countries like the Koreas, Taiwan, the United States and Canada. Mongolia should re-examine the validity and applicability of centralised planning systems, dictated by mentors from agricultural and industrialised economies: this includes the old Marxist-Leninists, but equally the IMF and World Bank strategic plans and the successful economic policies of Asia's 'Four Tigers'.

Another strategy lies in recognising that the best source of low-cost economic power is the animal husbandry sector. Resources should be allotted to this key economic component in order to stabilise the food supply, protect the environment, and quickly develop, without a massive influx of foreign capital, a potentially very profitable export product from the livestock industry. The nomadic lifestyle should be supported by domestic policies, not maligned or ignored. Reforms aimed at revitalising the livestock sector and promoting free trade policies have already played a part in Mongolia's history. In the 1930s the Mongolian government reversed the stringent collectivisation measures of the Leftist Deviation and implemented within five years a series of reforms which increased livestock numbers to levels still not repeated. This 'New Course', enacted under Prime Minister Gendun, must be analysed by Mongolian planners to see if some of the same programmes could be modified to suit present conditions. Mongolia does have a successful experience to use as a model; its economists are not faced with a vacuum when initiating policies to stimulate the nomadic economy.

Today Mongolia is part of a rapidly developing international economic system, which can only spur its growth. The first step down the road to economic development is to fully understand where Mongolia has come from, that is to comprehend the dynamics of the traditional nomadic economy. Underneath Mongolia's former command economy is not the remnant of a commercialised agricultural society, but a nomadic economic structure with its accompanying social and cultural values, still vibrantly alive. It is upon this foundation that the country's economic development towards privatisation, democratisation and modernisation must rest. The challenge for Mongolia is to find a new formula for sustainable economic development, which maximises the strengths of the traditional animal husbandry economy, but integrates this heritage with the free market economy. Such an economic strategy is more suitable than trying to graft an unnatural, expensive, Western-style agroindustrial economy onto a land with a harsh climate and rigorous geography, hostile to typical sedentary economic activities. The last decade of this century should find Mongolia evolving into a new society which still preserves its nomadic traditions and protects the ecology which nurtured it over the centuries. If Mongolia adopts such a strategy to guide its economic development, it will assuredly meet with great success.

Notes

1 '*Inter Nationes*: Conceptual Changes Will Lead to Growth', *Mongol Messenger*, July 28, 1994.
2 'Government Policy Affects Industrial Output Adversely', *Mongol Messenger*, July 28, 1994, p. 2.

Chapter 6

Lake Hovsgol – Selenge River Project, Mongolia

Zane G. Smith

Why Sustainable Development?

Dr Hal Salwasser of the University of Montana, USA, writing in the American *Journal of Forestry* in August 1994, provides a perspective for sustainable development. Quoting several sources, he describes the situation we find ourselves in as we attempt to survive on this planet. Perhaps of most significance, world population has increased elevenfold in the past 300 years, from 500 million to 5.5 billion. He points out that every living thing needs space and resources to survive, and the space available for each human being has been reduced to only one-ninth of that when the Industrial Revolution began. Every other living thing is thus is also being crowded, except perhaps those that benefit from human intervention. This, of course, has enormous implications for biodiversity and the health of the world's ecosystems.

The world's population has doubled in the last forty years, and while production has significantly increased, per capita availability of resources has declined. The gap between the affluent and the poor on this globe has doubled. Some will argue that population reduction is the only option. It certainly would be desirable for population to stabilise, but a more manageable option, I believe, would be to determine the best use of land and resources and then live with them in a sustainable way.

A Definition of Sustainable Development

The notion of sustainability was noted as early as the 18th century in the writings of Thomas Malthus. More recently the concept was specifically defined as the centrepiece of the Brundtland Report, commissioned by the United Nations in 1987. Briefly, sustainable development was defined as 'development that meets the needs of the present without compromising the ability of future generations to meet their own needs'.

The meaning of sustainable development has been debated ever since, but until now has failed to be implemented in any large-scale fashion. Much of the confusion has arisen because 'sustainable use', 'sustainable growth' and 'sustainable development' have been used interchangeably; but their meanings are not the same. 'Sustainable growth' is a contradiction in terms. Nothing physical can grow indefinitely. 'Sustainable use' is applicable only to renewable resources: using resources only at rates that do not exceed their capacity to replace themselves. Ecologically Sustainable Development, Inc. has defined 'sustainable development' as: 'improving the quality of human life while living within the carrying capacity of the supporting ecosystems'.

A 'sustainable economy' can occur as a result of sustainable development. Such an economy maintains its natural resource base, rather than consuming it. Development can continue by adapting, and through the application of improvements in knowledge, organisation, technical efficiency and wisdom.

World leaders began to coalesce around a commitment to the concept of sustainable development at the 1992 United Nations Conference on Environment and Development in Rio de Janeiro. *Agenda 21*, the document produced by the conference, promoted a strategy for sustainable development, but fell short of providing for its implementation on any kind of scale.

Lake Baikal and the Lake Hovsgol-Selenge River Programmes

Interestingly, the first large-scale planning and implementation of sustainable development occurred in Central Asia, in the drainage basin of Russia's Lake Baikal. It was the result of a cooperative effort led by Ecologically Sustainable Development, Inc. (ESD), an American non-profit organisation, working with both Russia and Mongolia. This enormous basin lies partly in Russia and partly in Mongolia, and covers 32 million hectares (an area the size of France and Belgium combined). The Mongolian portion of the basin includes the Selenge River Basin and Lake Hovsgol. Planning for the Russian portion was completed in March 1993 and implementation is now under way. The Mongolian half of the plan has been agreed to and is presently being published.

A similar project for the Ussuri River Basin in Northeast China and the Far East of Russia is midway through completion. An agreement to proceed with a sustainable development plan has also been signed by the government of Russia's Altai Republic, known as the 'roof of Siberia'. Bolivia, Chile and Nicaragua have shown interest, and a small project is being developed for lands controlled by a group of indigenous people in northern British Columbia, Canada.

We have now developed two models of sustainable development planning, which are complemented by our early experience of implementation. Let me now turn to the principles and activities which resulted in the preparation of these two landmark programmes, jointly developed by ESD and our Mongolian and Russian colleagues.

Assumptions

If sustainable development is to be achieved, there must be a willingness to place limits on how society can use the land and its resources. There is an assumption, then, that the land and resource base in a large planning area should be 'zoned'. In other words, land and resources should be divided into small units, which are categorised. The uses which the units of land and resources may be put to are defined by the sustainable development plan, to ensure that they are used in the best possible way. Thus, limitations are imposed on land use which take into account the land's inherent capabilities; the use of surrounding lands; the demand for resources; and the cultural norms of the region. Ideally, the limits on land use should reflect both ecological and socio-economic values.

A further assumption is that no one should have the right to use land in a way that will degrade the environment of others. Zoning and other policies affecting land-use should improve the ecological integrity of the land, preserve biological diversity and meet the cultural and economic needs of the people.

Listen to the Land – Listen to the People

Ecologically Sustainable Development, Inc., working with its Russian and Mongolian counterparts, adopted a theme of 'Listen to the Land and Listen to the People'. 'Listen to the Land' directed that we determine the inherent capabilities and limitations of the land and associated water resources. For most land there is a range of possible uses, but the use or uses which are selected must be sustainable, i.e.

the land must be capable of sustaining that use. Knowing and understanding land capability allows decision-makers to make the necessary choices to select uses so that the overall land-use pattern satisfies people's needs and those of biodiversity and healthy ecosystems.

'Listening to the People' acknowledges that human beings are also an essential part of our ecosystems and thus must be involved in decisions regarding their use. People have needs and desires that must be taken into account in land-use decisions. Sustainable development is only possible when people understand and participate in the choices which affect them. Every sustainable development plan needs to start with the education and commitment of public and private officials as well as the people themselves. It is then important to provide opportunities for public participation at every major step of the process.

The Process

The Lake Baikal and Hovsgol-Selenge River planning efforts, the Ussuri River and Altai projects have all followed a similar process. It is important to note that none of these programmes are American. They are and must be owned by the host country. Collaboration with Americans or other outsiders can strengthen the planning effort, but outsiders cannot unilaterally produce a sustainable development plan that will work and be accepted. It should be understood, too, that sustainable development is not an end point, nor a revolution. Rather, it is an evolutionary process that brings together interdisciplinary talent, NGOs, government and people, to analyse, develop and adopt a plan, and then implement, monitor and continuously improve it.

Thus, our process involves:

1 Agreeing with local government, scientists and residents on an area, timeframe and division of labour that will guide cooperative planning for sustainable development. A signed agreement is necessary, outlining objectives, responsibilities and a timeframe.

2 Collecting data on the physical and biological characteristics of the land and the needs and desires of the people.

The land data can usually be collated from existing sources. They should cover soils, water, topography, flora and fauna, climate and other relevant factors. Using these data, interdisciplinary teams categorise the land according to its resources and economic capability.

The social, political and economic data are collected through individual and group meetings with government, interest groups, communities and the general public. Consultation and public meetings have, in our experience, proved to be fertile sources of information throughout the planning process.

3 Based on analysis of both sets of data, zones of use are proposed. In our programmes, the proposed zones were depicted on 1:1,000,000 scale maps and presented for public review and comment. We have used the IUCN (International Union for the Conservation of Nature) categories for protected areas; the development categories we have used are arable land, pasture land, industrial land, resorts and recreation land, watershed protection areas of forest and steppe, managed forest lands, and cities and settlements. This consultation process should ensure that the zones adopted are consistent with both land capability and social priorities: in short, responsive to people's needs.

4 Prepare a draft Report to accompany the zone map. The Report describes the area and the context within which sustainable development is proposed. Each land-use zone is described, and preferred and conditional uses for it assigned. Performance standards are outlined for each use, based on the limitations of the land and resources and the technology which is expected to be available. Finally, recommendations for implementation are offered, including the legislation and administrative structures that will be necessary, monitoring and provision for more detailed local planning and making adjustments to the plan. An Appendix contains related issues and recommendations.

5 At this point a final public review of the draft documents is conducted. It is followed by consensus-building efforts by the interdisciplinary team and publication of the Report and Zoning Map.

6 After the programme has been published and distributed, the relevant levels of government consider and adopt it. In some cases revisions will be made at this stage to reflect the governments' own political priorities and realities. If the process we have used is followed, this step should be more or less automatic, since both legislative and executive branches of government will have participated in the decision-making already.

7 In most cases, legislation and regulations have to be passed to assure implementation. This is an extremely important step if the programme is to have any hope of being implemented. The

legislation must include provisions for financing and enforcement.

8 The next stage is implementation. Because the concepts and provisions involved may be unprecedented, it is desirable to establish demonstrations or model activities, which illustrate the overall direction of the programme. In many cases, government, with the help of international organisations, is willing to subsidise the cost of such projects.

9 Because sustainable development is a process, monitoring is essential. As implementation proceeds and demonstration projects reveal new information, the Zoning Map and the Report should be revised accordingly. Monitoring should involve the public, government and interest groups. The appropriateness of decisions, data, analysis, performance standards, zone boundaries, and enforcement all need to be considered. The programme is best viewed as tentative, with a willingness to 'try it, monitor it, fix it and try again'. This approach is often called 'adaptive management' and resembles a giant RD&A project (Research, Development and Application).

10 Specific projects and development activities such as forest management, resort construction, etc. will require more detailed planning. The sustainable development programme only defines what uses should be allowed in which areas, and sets performance standards which should guide development. More specific environmental assessments and detailed resource planning are required to fully implement sustainable development programmes.

Now and the Future

We believe the Russian and Mongolian programmes are the world's first two models of sustainable development planning on a large scale. It took vision, determination and persistence on the part of both countries to initiate these programmes and to stick with them through some extremely difficult times of transition. In particular, in Mongolia, Dr Z. Batjargal, Minister of Nature and Environment, provided inspiration, leadership and foresight as Chair of the Advisory Board for the Lake Hovsgol-Selenge River Project. Academician D. Baatar, President of the Mongolian Academy of Sciences, served as Vice-Chair of the Board. Dr G. Purevsern was the Mongolian Team Leader and Dr M. Badarch served as Project

Coordinator. Over 50 other Mongolian and American scientists and leaders served on the planning team and board. Their contributions will affect all the generations who follow.

The principles and concepts of sustainable development are similar the world over, but cultures and resources can vary greatly by country and region. Central Asia offers unique cultural values and resource circumstances. We believe that processes akin to that outlined above will lead to sustainable development which serves both the people and the land.

Plate 1: Tibetan nomads, Sichuan province

Plate 2: Tibetan nomads, Sichuan province

Plate 3: Tibetan settlers, Sichuan province

Plate 4: The Muslim minority in Qinghai

Plate 5: Gardeners at work, Qinghai

Plate 6: Local fishing, Angara river,
southern Siberia

Plate 7: Outskirts of Irkutsk with ring of dachas, southern Siberia

Plate 8: View of Lake Baikal

Plate 9: The Yellow River near Baotou, Inner Mongolia

Plate 10: Massive deforestation, Sichuan province

Plate 11: Trucks with Tibetan timber heading for Chinese factories

Plate 12: Water well, Mongolia

Plate 13: Nomads in northern Mongolia

Plate 14: Present day Ulaanbaatar, capital of Mongolia

Chapter 7

Environmental Sustainability, Development and Planning in Tibet

Provisional Findings and Conclusions

Graham Clarke

Background

This chapter is based on fieldwork carried out from Oxford University in Tibet, mainly in 1990 and 1991, together with the Government of the Tibet Autonomous Region of the People's Republic of China. This work involved the Economic and Planning Commission, the Commission for Agriculture and Animal Husbandry, and the Tibet Academy of Social Sciences. The research was carried out by Graham Clarke, Henry Osmaston, and Julian Wells from Oxford, together with various technical specialists and officers of the Tibet Government, as part of a wider programme of case-studies, advisory work and training, which also looked at problems of urban development with the Government of Lhasa Municipality. This latter part of the project was designed to include assistance in project planning, formulation of key issues and corresponding sector programme planning, and the formulation of general environmental guidelines.[1]

The chapter follows in the main part from case-studies on the impact and processes of development and recent change on the economy, society and environment. There was a 'three by two' design for six such case studies, the three dimensions being dominant pastoral, grain producing and forestry producing areas, and the two dimensions, locations close and distant from developing infrastructure. The intention here is not to present the detailed results of the case-studies, which will be published elsewhere; rather, the intention is to raise policy-related issues, namely our overall findings and conclusions as they refer to local processes that result from economic development, their impact on non-renewable natural resources, that is environmental sustainability, and on planning.

The case-studies were carried out in the large central region known as Lhasa Municipality, which is the size of a prefecture and stretches up to include part of the Tibet Plateau, and have been illuminated by other periods of research in culturally Tibetan areas of the Himalaya, in the Province of Qinghai, and through advisory work for bilateral and multilateral agencies in central Tibet. The overall areas looked at range from Nyelam in the west to Damxung and Reting in the north to Mozhugongkar in the east to close to the Sikkim border in the south. Two of the main case studies have been at locations in Damxung and Quxu (also commonly spelt Damshung and Chushu) counties. One of these is largely a livestock-producing and the other largely a grain-producing county.

The Issues

The first issue is the way that economic growth is linked to pressure on natural resources. The traditional and current base of the economy depends largely on natural, biological, resources: pastoralism and agriculture together form more than three-quarters of the Tibetan economy.[2] Since 1980, following on from economic reform, there has been an increase in production of livestock and grain as commodities for markets, reflecting an increased overall regional demand for material consumption. There has been an associated increase in the numbers of livestock at any one time on the grasslands. This production in the rural areas has been linked to market sales in the growing towns. Over the same period there has been an increase in the numbers of people resident in the few urban central towns.

This reflects a more general point, namely that development normally is conceived of as economic development and in practice is associated with a growth of commodities and services available in urban areas. It is this growth in livestock and agriculture production linked to urban demand and trade that has created a pressure on natural resources. In particular, the pressure has been on biomass (vegetation and animal products) that, due to the particular combination of topographical and climatic circumstances in Tibet, is in part likely to be non-renewable.

The household, the major local institution of production and consumption, uses all these items, and the use of each will affect the use of all the others. As elsewhere in the Chinese polity, there has been a 'springback' effect from economic reforms after a long period of containment, with a general increase in motivation from the

126

introduction of the individual responsibility or household contract system. I have covered the effects of economic reform on pastoralism elsewhere;[3] the recent increased prosperity of some grain-producing villages close to urban centres is due to an increase in production at the household level. The main factors involved may be summarised as follows:

a) modest increases in the yield and reliability of the crops, with limited farm-mechanisation and use of fertiliser;
b) use of this farm machinery for the market-sideline activities that now are possible under the economic reforms;
c) the availability of rural capital under the economic reforms; the purchase of this machinery has often been financed by official or private loans, though sometimes by internal savings.

Generally, economic growth in peri-urban grain-producing rural areas comes both from an increase in grain production, and from a central and urban subsidy. The ownership of transport has been linked to the availability of capital loans for the purchase of machinery (tractors and trucks) in rural areas. This has been accompanied by temporary migration and remittances from the sale of services for labour and transport in the town, particularly in the construction sector. This economic growth through sidelines is indicated in the example of Chushu, and that of sale of grain in Pa-nam (also spelt Bai-nang), between Gyantse and Shigatse.

The second issue is why there should have been this dependence on biomass. There are of course historical reasons for initiatives in the agriculture sector, but there has been a more general constraint to other forms of economic development in Tibet, namely the shortage of alternative sources of energy. This shortage has been reflected in both patterns of investment and sources of domestic consumption, and growth in agriculture and livestock production has created additional pressures on biomass.

This demand has been localised close to roads and peri-urban areas. There have been past technological initiatives here, such as the use of thermal springs to the north of Lhasa for the generation of electricity, with the help of UNDP New Zealand, and latterly equipment from Italy and Israel; in 1990 these were supplying up to 40% of Lhasa city's electricity. There are also the current projects for further electric production through hydro-electric power under the 'One River – Two Streams' programme, linked to four main constructions in central Tibet, a point to which we will return below.

However at present there is an absence of inexpensive, sustainable, energy supply from other sources and, for the main part, energy depends directly on biomass. This includes wood, bushes, the pasture itself, and dung, as fuel; and the pasture, bushes, and grain for direct consumption as food and fodder, and for onward resale and consumption. The systemic, cumulative effect of these pressures is not sustainable and can result in land degradation.

Land and Natural Resource Degradation

The effects of increased demand on each of these items is not the same. Grain and dung are of course renewable natural resources. Pasture, whether plateau or hill, and forest, are in principle renewable but in practice less so. The Tibetan high-altitude ecosystem has a certain type of fragility and stress not encountered in irrigated, climatically moderate, lowlands. Degradation of natural resources in Tibet particularly affects these pastures and primary forest cover.

Agriculture and Livestock

There is a great difference between the valley areas of the Yarlung-Tsangbo, and the Kyu Chu and Nye Chu (also spelt Ka-chu and Nyang-chu), with relatively deep deposits of fertile alluvial soils and good irrigation, where such increased grain production is possible, and the Tibetan plateau and the hill areas above these river valleys, which have far a more fragile ecological system with thin soils.

The economically more developed farming areas, as for example in the peri-urban areas of central Tibet close to towns or with reasonable road access, have some areas at risk of degradation to the natural environment. This tends to be not in the main areas of irrigated grain production themselves, which apart from the continued natural risk of the encroachment of 'vertical deserts' of drifting sand dunes, are much as they always have been. In the areas between Gyantse and Pa-nam, the probability of rapid and irreversible soil erosion is low. In some areas that adjoin the Lhasa river, small amounts of land are being lost in two ways. Floods from eroding tributary catchments spread rocks and debris on cultivated fields, and the banks along the steeper reaches of this river are becoming eroded. Along with the shifting sand dunes, these processes are perennial, natural, and almost inevitable at such sites: but they are relatively minor problems.

These fertile river valleys with their deep soils, for all their agricultural and human importance, form under 1% of the ground cover of central Tibet. Over much of the rest of the central region the topsoil is a wind-deposited loess that though fertile, gives a far thinner hill cover. The main man-made risk of erosion comes from increasing the pressure on the more marginal dryland hill areas in the production system for grazing, and here a loss of cultivable area and soil is likely. Historically, this topsoil was protected from erosion by a naturally dense and productive turf, and was used mainly for extensive grazing of livestock in a transhumant fashion, the annual rotation of which evened out the pressure on the land over time. In a significant number of central areas it is this land, whether marginal hill-land or customary winter-grazing, which now is at risk.

Both on the edge of the plateau and in hill areas, especially those peri-urban locations close to roads that are *en route* to market, recent increases in livestock numbers and uncontrolled grazing have caused local but extending destruction of the turf, erosion of the topsoil, and exposure of the relatively infertile subsoil. Though not all gullying results from man's activity and much is natural, new gullying has also started, down to the bedrock. The process of degeneration is likely to accelerate as the remaining land available to grazing diminishes, and regeneration of the productive capacity of this grazing land will be very difficult if this process continues.

On the highland Tibetan plateau itself, there also are problems with pests and weeds on pasture. Burrowing rodents (mainly pika but also marmots) are pests that damage unirrigated pasture in large areas of Tibet; their populations show large natural oscillations, control is difficult, and the role of natural predators in keeping the population under control may have been diminished as their numbers have been controlled to safeguard increased numbers of grazing cattle. The invasion of degraded pastures by non-nutritious and poisonous weeds has also begun to present a serious problem.

Forest

It may have been the case in the past that the small areas of natural forest in central Tibet were at risk from over-felling, and also from the free grazing of livestock which prevents regeneration. These are not problems at the present time, and old trees are protected in central areas. The state has been quite aware of the advantages of planting new 'shelter-belts' of trees in grain-producing areas to protect fields against wind-blown sand and erosion, and as an alternative supply of energy to existing biomass. The planting of these is encouraged. In addition, though in formal terms control devolves on the individual, collective control at village level has been maintained over the felling of trees during the economic reforms, and there is a strong communal ethic to the extent that such shelter belts have many of the aspects of common-property resources with a collective sense of responsibility and make private unplanned felling of these trees unlikely.

We have as yet no direct information on forestry in the south-eastern valleys off the edge of the Tibetan plateau, such as at Medog and Nyinje (also spelt Metog and Nying-tri). However, following the improvement of the road system and expansion of trucking enterprises, it is clear that much wood as fuel and timber for construction is being brought to central Tibet, and part of it comes from this area. Various informal reports and comments on forestry in the south-east of Tibet and adjacent highland areas of Gansu, Yunnan and Sichuan are available. They imply that timber is exported directly from the south-east by river and truck; that these regions are supplying fuel and timber at a rate that may well exceed a sustainable yield. More general comparative evidence suggests that wasteful clear-felling in steep-sided valleys and on the plateau can promote soil erosion, flooding and river siltation. In this south-eastern highland area it is clear that the erosion, altitude, and local climate would act to inhibit natural regeneration. To some degree, this use of timber from further afield may act to decrease the risk of natural degradation from over-exploitation in central Tibet; however, this may merely transfer the problem elsewhere, and postpone some of the consequences of environmental degradation in central Tibet.

Hence the pattern for forestry appears as the reverse of those for pasture and dry hill land: whereas in the former the pressures are in the areas remote from the key central areas of economic development, in the latter the pressures are in the central areas; both are derived from urban and market demand.

Underlying Processes

The increased demand on biomass is both for immediate consumption by rural and urban households, and for production for market and economic investment. One key factor has been the partial sedentarisation of traditionally transhumant nomads near roads, which is where the best grazing and transport tends to be located:

this has increased the pressure on both prime-quality grassland and firewood.

As well as the increased demand for outputs such as timber for construction and firewood, the increased demand for energy also removes biomass more directly from early parts of the input cycle to the land. Traditionally, the fertility of the pasture area and fields was maintained by applying manure; but now the shortage of wood-fuel in some areas has resulted in less manure being put on the fields and lower amounts of dung left on the pasture, as it is used for fuel. This non-virtuous cycle serves to accelerate the process of pasture degradation.

The studies suggest the following process is operating in central areas with access to roads. Households which were formerly transhumant come to spend more time at tents at locations which formerly were only seasonal pasture. These tents become houses, and then the principal settlement of the extended family group. This sedentarisation, together with increased numbers of livestock, leads to a general pressure on non-renewable timber and shrubs, and the pasture itself for use both as grazing and fuel; in turn, this leads to the increased collection of animal dung for use as fuel, so removing it from the traditional use as fertiliser on the grassland. The focus to this process tends to be along the road areas, as other households from more distant settlements also bring their livestock through these nodal locations to market, or place them with relatives to fatten before their final journey to market.

This process of pressure on biomass from economic growth is accentuated by the growing direct energy demands of towns. Pasture has always been cut up by nomads for use in tent construction and as fuel and now is also being removed for market sale for fuel as, for example, in Rinbo County (also spelt Rinbu) north-east of Gyantse.

Hence the rate of utilisation of the pasture is no longer determined by the demands of a static population of livestock equalised in pressure across the area by an extensive transhumant cycle of grazing and settlement over the course of the year, which allows fallow periods; increasingly it is being determined by a growing economic and specifically urban demand, markets, and the location of roads. This local specificity implies that the problem is not widespread; yet the intensity makes for a negative environmental impact in those highly visible locations, which combined with the thin soils, and climatic extremes, may make some changes towards erosion irreversible.

In these local peri-urban areas there can be a progressive downward spiral locally of the condition of the pasture and rough grazing. In terms of the grassland as a whole this may not be important as such locations make up less than 5% of the total available in Tibet; similarly there is no lack of rough hill land in Tibet, much of which in any case is semi-eroded. However, the land degradation is occurring in the key peri-urban and road areas where land is needed as a resource for the livestock industry. Furthermore, the extent of this land degradation is likely to increase rather than stabilise, both to substitute for newly non-productive land, and as the road network and regional economic integration increase.

Administrative Perception and Response to Impact

In Tibet, problems of grassland quality and land degradation from the pressure of pastoralism have been recognised by government at all levels for the peri-urban areas earmarked for pastoral production, and various official measures have been proposed for their solution. One such set of measures has concerned the attempted use of technology directly to control the land.

In trial areas herbicides have been used to destroy non-nutritious grasses and weeds, and inorganic fertiliser has been applied to increase yields. Compensation through the use of imported inorganic fertilisers is useful and can increase yields; but it is unlikely to be a long-term substitute as there is a decrease in organic content and close-grazing can still result in exposure of the soil and erosion. Care has to be taken to avoid the direct human dangers of herbicides, and the long-term problems of such large-scale chemical control are not yet fully recognised. Beyond these facts there are economic factors: there is a subsidy through the transport sector for such chemicals which are costed at the same price as at landing on the eastern seaboard of China. It is not at all clear that the economic benefits exceed the costs at market prices.

Another favoured approach involves control of the pasture to exclude animals at certain seasons, and to help regulate ownership. Fencing is one such measure, and the policing of ownership is another. Again, the costs of such fencing may be high, but here at least it should be a capital expenditure rather than a recurrent cost, as with fertiliser. However, regulation through a specialised police group is unlikely to be cost-effective in rural areas, due to the vast distances and low population densities that make such administrative control here, as with other services, prohibitive.[4]

Fences can easily be cut; measures such as fencing can only be successful when they are proposed and come from the communities themselves, who have to take active responsibility for their maintenance and policing. Ultimately, such fencing shades over into private ownership and it may well be that such household ownership is the form of ownership which society is moving towards; however, in between there may be other stable states of common property management, and fencing may also have a role here. There may well be a rationale for fencing in areas of grassland with higher population density, and through traffic, closer to roads; there is unlikely to be a rationale for fencing of high, marginal hill or mountain areas. However, the key factor is the initiative of the local community itself and the technical measure of fencing alone cannot address the problem.

There is some local participation in implementation; but there is little in planning or policy formulation, and it is only now that local rights to renewable natural resources are beginning to be seen as secure. To summarise, the wider comparative experience indicates that measures for conservation control by external policing and investment in material technology alone are expensive and may be ineffectual. They are especially likely to be inappropriate in areas of low population density, if for no other reason than because of the high per unit cost, and attempts to implement them may themselves accelerate the problem of non-renewable resource exploitation.

Counties are classified administratively into grain- and livestock-producing counties, the agriculture department being responsible for the former, and the livestock department for the latter. The widespread hill erosion is from overgrazing on the hillsides of grain-growing counties in peri-urban areas. Problems of land degradation from the pressure of livestock are not well-recognised in counties registered as primary grain production areas: these areas are not recognised as grassland and there are few measures, other than plans for fencing by the state, which have been proposed for regulation. The hillsides where the degradation is taking place are high above the grain-producing river valleys which concern the agriculture department, and are beyond the province of the livestock department. At the same time, the increase in production under economic reforms, through predominantly diversified mixed agro-pastoral households, makes for a parallel pressure through both sub-sectors.

One key problem here is the absence of any economic valuation of this hill land as a natural resource, other than through the marginal gains in production of bringing it into the economic system by livestock grazing, which can degrade it. These areas need not be fenced off into reserves, or over-utilised under a crude market valuation; but they need to be valued by the state under a sophisticated 'environmental economics', with local common property regimes for their management.

In forestry, it is encouraging that there are plans for afforestation of uncultivated valley land in central areas. As well as funds for investment, effective planning and implementation also requires an understanding of economic, institutional and community needs and priorities for access to this resource, and the Tibet government seems to be well aware of these needs and the priority of reducing pressure on other biomass for energy. However, it is not at all clear that such policies are applied in the more remote areas. Again, natural resources such as primary forest cover and vegetation are not valued properly as assets, and hence clear-felling does not result in a loss in capital assets. Comparative evidence suggests that few local governments, moving towards a market system from a centrally subsidised state system and hence with reduced capital, can resist the windfall gains to be achieved from the sale of timber when roads make these areas accessible.

There is a high awareness of problems of urban pollution by the Lhasa City administration. However, urban growth in the capital has not been accompanied by a sufficient investment in urban services, such as sewerage and sanitation, solid waste disposal, or hygienic abattoirs. Indigenous popular attitudes, whether Tibetan or Han Chinese, are not attuned to problems of urban pollution.

Traditional Tibetan culture teaches a respect for all life, and may be of great help in planning popular initiatives in social and environmental sectors. Yet though some traditional ideologies can be maintained in the face of economic change, comparative evidence suggests change often operates not so much by traditional ideologies carrying over into the present, as by the popular penetration of new, exogenous, ideas which encompass them, leaving them eventually as a 'folk' residue. Local beliefs are likely to change further, if unevenly. Neither should it be presumed that all traditional ideologies are necessarily beneficial, for they can lead to unexpected problems. For example, there is the problem of hygiene with the large numbers of dogs in Lhasa City. The human population in 1990 was over 120,000, while the city was estimated by some officials of the Lhasa Municipality to contain upwards of 15,000 dogs, and there was some popular resistance to having nightly marauding dog packs destroyed.

Overall, there is an institutional awareness of the problems of natural environmental degradation, especially in as far as they impact on problems in urban services and pollution, within the Tibet government. However, in 1991, this awareness of problems had not yet led to a clear, programmatic, prioritisation of goals, assessment of environmental costs, integration of environmental guidelines in economic planning, policy on taxation or other controls. Environmental planning was carried out primarily by a separate bureau which had an information recording and support, rather than a planning, remit.

Policy Issues and Planning

Currently, there is a new emphasis on local assembly industry for export in central areas of Tibet, and trade in commodities with neighbours in India and Nepal. However, one of the earlier factors in the regional economic strategy has been the establishment of infrastructure for access to natural resources for export from the region, which was a general strategy advocated under the eighth five year plan for western regions of China.

Though the livestock caravans still carry out an intra-regional trade between plateau and 'lowland' regions, there has been a growth in formal and informal market activity. In part this is linked to a modern renaissance of long-distance trade to Qinghai, Gansu and Sichuan and the spectacular growth in the trucking sector. This has been an important contributory factor to the peri-urban economic growth in the livestock sector, in the linked trade in salt and grain, and the delivery of timber for market, and this growth in the transport network has acted to increase the degree of regional market integration.

Here, a subsidy for transport and machinery, and accordingly for the intensification of agriculture and the construction industry, is one of the critical factors behind the present increase in economic output in both rural and urban areas. Key features of this subsidy have been cheap fuel for road transport and the availability of capital through loans for investment, and low initial prices for equipment that do not reflect the costs of industrial production or market prices.[5]

Though an import substitution programme for grain production is planned by the state, transport subsidies for fertiliser and other inputs remain, as they do in other sectors, both for immediate consumption, and for production for market and investment. As has been pointed out above there are heavy state subsidies to these fertilisers in terms of transport from the eastern seaboard of China, which are not taken into account in a cost-benefit analysis, despite being transported for thousands of miles across China.

Stopping the supply of chemical fertiliser would result in an immediate reduction in grain production; yet a dependence on the import of cheap oil (piped from Golmud), fertiliser (trucked from Golmud and Xining) and consumer goods is not a sustainable long-term economic strategy. Subsidised imports, and low valuation of non-renewable natural resources, can only be a stop-gap to allow internal reconstruction and industrial growth. The removal of subsidies on manufactured goods and the proper valuation of non-renewable resources may give rise to transitional problems, similar to those encountered in Mongolia without the Soviet fuel subsidy. Other countries have had to undergo structural adjustment programmes; in Mongolia there is often an inflation not just in the prices of consumer goods, but in the cost of materials necessary for industrial production, which affects the modernising market sector rather than the traditional subsistence and barter areas. Yet, the creation of a modern market sector without an overall dependence on external subsidy reduces long-term economic risks.

Economically sustainable production, namely production that is competitive at market prices, also may promote the conservation of the environment, in that the removal of transport subsidies will limit subsidised over-exploitation of the natural resource base, and limit economic enterprises to those which can cover their own recurrent costs from profits, that is, be sustainable.

As a counterpoint to the linked issues of market factors and subsidies, there also is the new ideology of 'environmental economics' which we have referred to above. One factor allowing environmental degradation in areas open to economic development has been treating natural products such as salt, grassland, and forests as 'free goods'; that is, their zero costing in economic calculations for production (including opportunity costs of short-term consumption), and the absence of any investment in long-term renewal.

A further factor that is lacking is the control of local demand and consumption through indirect taxation. For example, the zero valuation of empty bottles without deposits in Tibet, leads to the dumping and smashing of glass bottles in the town and countryside.

This may well be economically 'rational', because the land on which they are disposed of, for all its natural beauty, is not valued financially as an asset.[6] Certainly, the presence of a bottling plant would be of help in valuing these as a renewable rather than disposable resource; but there could be further conscious use of pricing and deposits to value these goods and preserve the natural environment from this garbage with all its dangers to livestock and humans alike, and a negative impact on tourism.

The importance of allowing for local conditions in planning is widely acknowledged within the regional government. There is an awareness of the need to integrate local production and consumption, and to promote the renaissance of trade with India through the special trade zones in the central areas, eastern and western borders, and for border populations. Since 1991 the encouragement of border trade has benefited these groups.

At the same time, overall, planning often continues to be based on physical output targets and technical intervention measures intro- duced from lowland areas of China, without always making a full assessment of their technical appropriateness to ecological conditions on the plateau, or of the recurrent running costs involved. The effect of such standard economic interventions on a sparsely populated high-altitude ecosystem is not the same as on densely populated irrigated lowlands.

Energy

One key underlying constraint for balanced development in the region is sustainable local energy production, both for consumption and for light industry. The presence of this would lower the demand for increased energy on biomass and the land which, under the stimulus of transport subsidies, is literally fuelling economic growth. Any overall strategy for investment has to address this issue.

Elsewhere within the polity of lowland China there has been good work in compensating for the damage to forests of the 'Great Leap Forward', through the present policy for reforestation; but in highland areas more of the primary biomass is non-renewable, and to rely, alone, on secondary remedial strategies after implementing a programme of growth, may generate additional risks. Programmes of secondary afforestation are all very well, but do not solve the problem of degradation of non-renewable primary natural re- sources.

In Tibet, as in other highland areas of Central Asia with extreme natural conditions and a high degree of climatic stress, various aspects of the ecosystem are more fragile, and not all standard economic and service innovations are financially sustainable. The point is to make sure that unplanned, irreversible, changes to the non-renewable natural resource base do not come through the adoption of exogenous models, applied without assessing their appropriateness for the conditions at hand.

Common property resource management may be viable in traditional groups, that is those with a communal, a *gemeinschaft*, ethic. Private property regimes are viable in a more developed, commodity-based market economy, with a secure legal base in individual property rights backed up by civil law. However, between the two chaos may reign. If one relies solely on reactive demand-led development, by the time the demand has been translated into programmes, plans, and operating systems, and alternative sources of energy are available, some damage from increased demand on biomass for energy production will already have taken place. Hence even if a programme were introduced now there would still be environmental problems due to the time-lag in transition from biomass to other energy-generating processes.

There is a key programme of centrally directed agricultural investment in Tibet: the 'One River – Two Streams' Programme of Projects. As indicated at the beginning of this chapter, this is not just an increase in agricultural production but centres on irrigation and an increased production of electricity through hydroelectric power. As well as the thermal springs north of Lhasa there are four new structures proposed or under various stages of construction. Two are built already, one north-west of Gyantse, and the other the head pumping station from the Yarlung Tsangpo up to the Yamdrok Lake, which has attracted some international attention. Neither are fully operational, however. The others are a dam on a southern-side river at Pa-nam, and in principle a dam across the Tsangpo itself. Such systems have a hierarchy of functions: one is urban electricity supply, the next is irrigation of land for additional crops, and the third is embankment and flooding control. The first two factors can conflict with water allocation when there are low reservoir levels: in most parts of the world, supply of electricity to a modern growing urban consumer market and light industry wins out over the delivery of water for rural irrigation.

In considering the environmental impact of this hydro-electric power system, it is important to realise, whatever one's aesthetic feelings on the Yamdrok Lake, that hydro-electricity offers the main alternative to the use of less-renewable resources such as timber, shrubs and pasture, that is biomass, as sources of energy. There may still be a question of the appropriate scale and location of the schemes: large dams are not always the most efficient economic configuration; they are prone to siltation, and the World Bank has been reevaluating the life-span and environmental impact of such schemes; the benefits from smaller structures in agricultural areas, tied to the rural townships or villages and primarily under their control for irrigation and small-scale hydro-electric power, need also to be considered and *prima facie*, may be more capable of raising local living standards more directly. Overall, given urban growth, power supply from integrated hydro-electric schemes is a key feature to environmental improvement and economic development.

Notes

1 Assistance was given to the Tibet government in training for writing project proposals, through helping them redraft and consider their priorities for some twenty such proposals they had under prepararation in 1991.

2 77.4% of Total Regional Output by Value in 1989; this figure is calculated from data given in the *People's Daily* on 21st July 1990.

3 G. E. Clarke, *China's Reforms of Tibet, and their Effects on Pastoralism*, IDS Discussion Paper No. 237, Brighton, UK, 1987; reprinted in *Kailash: a Journal of Himalayan Studies*, Vol. XIV, nos 1–2, pp. 63–131, 1988.

4 The population density in Tibet is 0.61 persons per sq km, as contrasted to the overall China figure of 105 persons per sq km (Clarke, *loc. cit.*).

5 The comment of one local official, when asked if any cost-benefit calculations had been done for agricultural intensification, based on the market costs of fertiliser and fuel rather than the subsidised costs, is quite illuminating here. His reply was 'If we used market prices, then there would be no economic investment or growth in Tibet'.

6 Picnic spots and hilltops are such sites where bottles are left smashed, as much by Tibetans as by others.

Chapter 8

Desertification in Western China

Wang Tao

Introduction

Western China has abundant natural resources and a great potential for sustainable development. However, most regions in Western China have been seriously affected by land degradation, particularly desertification, which has become a major factor restricting economic and social development. On the basis of research and practice over many years, we believe that the desertification is caused by wind erosion, water erosion and salinisation, which result mainly from adverse human impacts. Desertification leads to a rapid decline in the land's production of biomass and potential productivity, and even to absolute loss of land resources. This creates problems not only in the natural environment, but in the economic and social spheres. The desertification situation is becoming more and more severe in Western China. However, some good examples have proved that the measures adopted to conserve land are feasible, and the process of desertification has effectively been controlled in some typical areas. The productivity of cultivated land and rangeland have gradually been restored.

The State of Desertification in China

Desertification is a serious economic and environmental problem which many countries are now facing, and which China suffers from in a severe form. For example, the area of land in Northern China which has been desertified through wind erosion, also known as sandy desertification, increased from 137,000 km^2 in the 1950s to 176,000 km^2 in the mid-1970s (Table 1), and continued to spread, reaching 197,000 km^2 by the late 1980s (Table 2).

Table 1: Distribution of desertified land in Western China (1970s)

Unit: km² Regions	Total area	Different types of desertified land		
		On-going	Severe	Most severe
Hulun Buir	3,799	3,481	275	43
Lower part of Nenjiang	3,564	3,286	278	
West Jilin	3,374	3,225	149	
East fringe of Da Hinggan Ling Mt.	2,335	2,275	60	
Horqin Steppe (Jirem Prefecture)	21,567	16,587	3,805	1,175
Northwest of Liaoning Province	1,200	1,088	112	
Xar Mulun River (Former Zhao Uda Prefecture)	7,475	3,975	1,875	1,625
North parts of Weichang, Fengning County, Hebei Province	1,164	782	382	
North of Zhangjiakou, Hebei Province	5,965	5,917	48	
Xilin Gol and Qahar Steppes	16,862	8,587	7,200	1,075
Ulan Qab Prefecture, Inner Mongolia (Beyond Daqing Mt.)	3,867	3,837	30	
Ulan Qab Prefecture, Inner Mongolia (Daqing Mt. area)	784	256	320	208
Northwest of Shanxi Province	52	52		
North of Shanxi Province	21,686	8,912	4,590	8,184
Ordos Plateau (Ih Ju and Ulan Qab Prefectures)	22,320	8,088	5,384	8,848
North of Ulan Buh Desert and the Great Bay of the Yellow River	2,432	512	912	1,008
North of Langshan Mt.	2,174	414	1,424	336
Centre and southeast of Ningxia	7,687	3,262	3,289	1,136
West piedmont plain of Helan Mt.	1,888	632	1,256	
South edge of Tengger Desert	640		640	
Lower reach of Ruoshui River	3,480	344	2,848	288
Centre of Alxa Prefecture	2,600	392	2,208	
Oasis edges in Hexi Corridor	4,656	560	2,272	1,824
Piedmont plain in Qaidam Basin	4,400	1,136	1,824	1,440
Gurban Tunggut Desert edge	6,248	952	5,296	
Taklamakan Desert edge	24,223	2,408	14,200	7,615
TOTAL	176,442	80,960	60,677	34,805

141

Table 2: Development of sandy desertification in China (1970s to 1980s)

Monitored region	Area (sq km)	Period of study	Percentage of land desertified:	
			in 1970s	*in 1980s*
Rainfed croplands in Bashang	17,250	1975–87	14.6	26.7
Rainfed croplands and grasslands in Chanar steppe, Inner Mongolia	9,056	1975–87	31.4	66.2
Upper reaches of West Liache River and Horqin sandy lands, Inner Mongolia	42,300	1976–88	68.5	77.7
Rainfed croplands in Houshan steppe, Inner Mongolia	46,660	1975–87	22.5	39.1
Ordos steppe in Ikh Cho League, Inner Mongolia	55,874	1977–86	80.1	85.6
Regions along the Great Wall in North Shanxi Province	18,046	1977–86	43.3	45.3
Shugui Lanke region in middle part of Alxa desert	1,573	1974–84	74.5	83.2
Lower reaches of Ruoshui River in western part of Aixa desert, Inner Mongolia	16,200	1975–86	21.5	36.8
Piedmont plain of Kunlun Mt. in Qaidam Basin, Qinghai Province	7,920	1976–86	55.6	70.4
TOTAL	214,879		49.5	60.7

The sandy desertified lands mainly fall into the following categories:

a) interlocking areas between cultivated and grazing lands in the semi-arid zone (about 40.5% of sandy desertified land);
b) undulating desert steppes in the semi-arid zone (about 36.5%);
c) marginal oases and lower reaches of interior rivers in the arid zone (about 23%).[1]

Most sandy desertified land is in Gansu Province and the Autonomous Regions of Inner Mongolia, Ningxia and Xinjiang. According to the results of research in the Goughe Basin of Qinghai and in central Tibet, sandy desertified land had reached 12,670 km^2 and 1,860 km^2 respectively by the late 1980s. Thus the total area of sandy desertified land in Western China is about 211,530 km^2.

The Process of Desertification

The main mechanism of sandy desertification is wind erosion after vegetation has been destroyed. When the wind erodes farmland and rangeland, the topsoil is carried away. Shifting sands accumulate on the leeward side and gradually develop into dense, mobile dunes. The original features of the surface are replaced by roughness, terraces erode and defoliation accelerates. These processes destroy the structure of the soil, which causes a rapid decline in biomass production and the potential productivity of the land, and even the absolute loss of land resources. In addition to wind erosion, water erosion and salinisation can cause desertification. In Sichuan and Yunnan Provinces, desertification is exacerbated by water erosion. Salinisation has occurred mainly in the continental river basin in Xinjiang, because of the misuse of water resources.

These processes of desertification are encouraged by human activities, which interact negatively with natural processes. Table 5 shows the dynamics of desertification in Western China, demonstrating how vegetation destruction caused by intensive land use is the breakthrough point for desertification.

Table 4 shows the relative importance of different human causes of sandy desertification in China.[2]

Strategies and Measures for Controlling Desertification

Over the last four decades, the Chinese government has paid a great deal of attention to combatting desertification, and this has resulted in expansion being checked in about 12% of cases of desertified land and in the rehabilitation of 10% of cases.

The task of prevention and control of desertification is a very arduous and complex one, requiring an enormous amount of work. Strategies to control desertification should be governed by the principle of uniting economic and environmental benefits: exploitation and use of resources should be combined with the control and protection of the environment. Strategies should take into account the natural resources and socio-economic characteristics of the desertified area, the problems of land utilisation which are causing desertification, and experience of combatting desertification in similar areas. Desertification control strategies should include the following main elements:

Table 3: Factors leading to Desertification

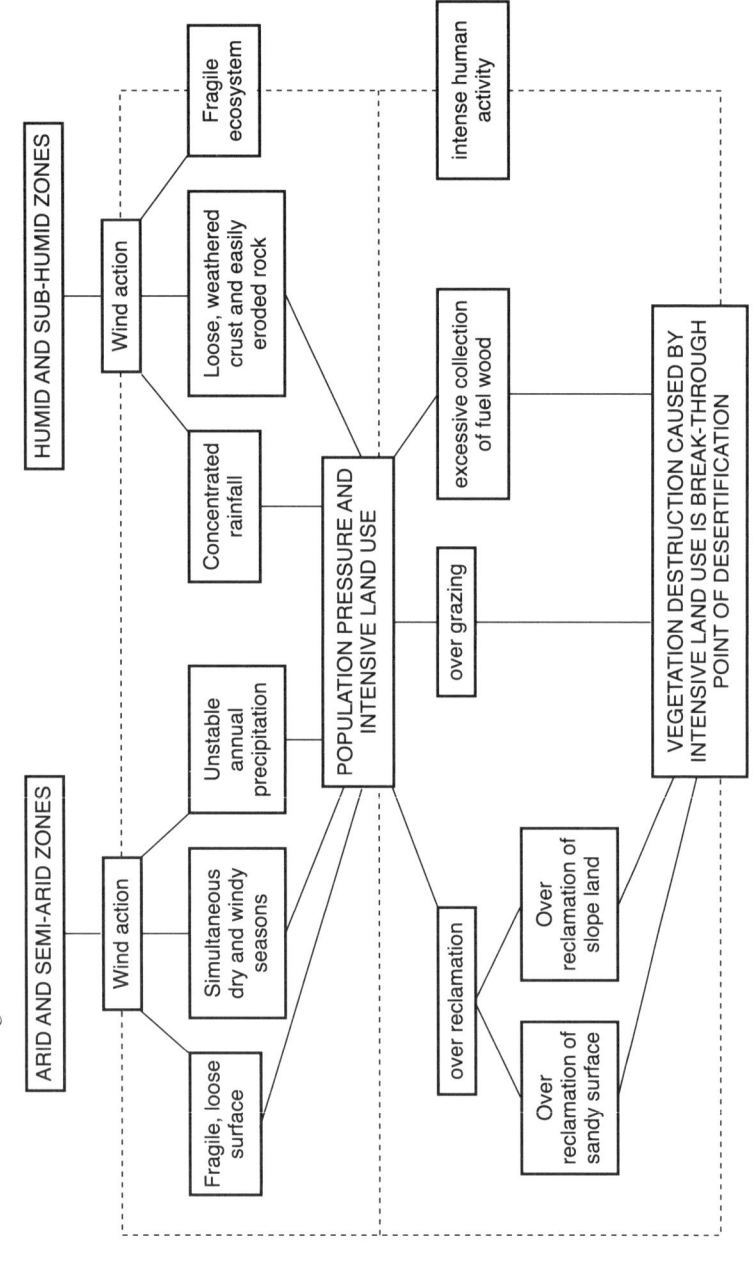

144

Table 4: Different human causes of sandy desertification in China

Causes of desertification	Percentage of the total area of sandy desertification
Overcultivation on the steppes	25.4
Overgrazing on the steppes	28.3
Overcollection of firewood	31.8
Misuse of water resources	8.3
Total	93.8

a) developing animal husbandry and forestry properly and limiting them so as to make reclamation possible;

b) controlling the growth of population effectively and reducing the pressure of human population on desertified land;

c) reform of the rural energy structure, especially the use of firewood for cooking and heating.

Observing these strategies, we can develop concrete measures to reclaim desertified farmland and rangeland.

Fencing

As pointed out above, sandy desertification occurs and spreads due to a combination of the fragility of the ecosystem and excessive human economic activity; thus, in a semi-arid climate, as soon as the human interruptions of excessive economic activity stop, the process of desertification gradually ends, and the land naturally returns to its former condition. Fencing is an effective way to limit the spread of desertified land, which has been adopted widely in semi-arid areas. After land has been fenced off for two years, both plant cover and biomass increase.

Regulation of Land Use Patterns in Marginal Areas

This policy involves converting land on the margins between farming and pastoral areas from dry-farming grain production to grazing and woodland. This achieves a twofold advantage: environmental and economic. Apart from control of desertification, such regulation enables flat fields with better soil and water conditions in basins or along river valleys to be more intensively managed. The success of the policy can be exemplified by some measures implemented on sandy

Table 5: Process of Desertification

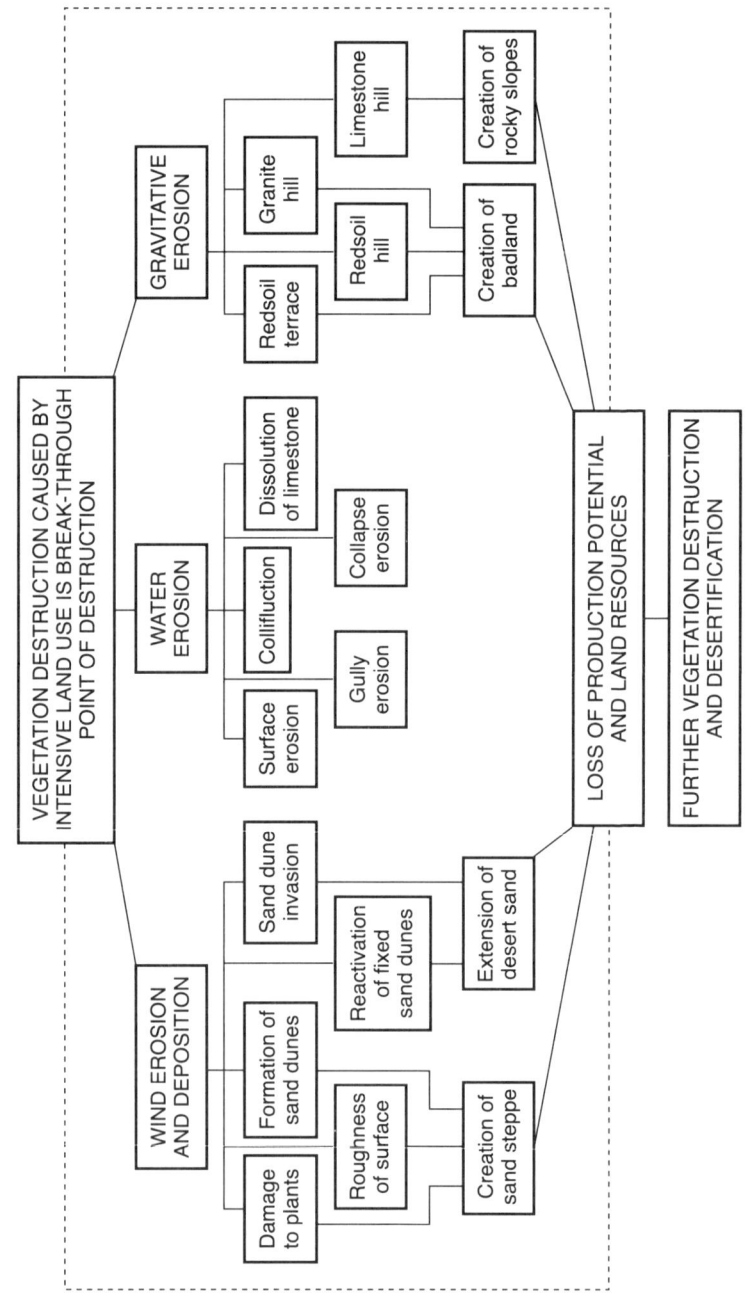

146

desertified land. For example, the Huanghuatala village in Naiman County, Inner Mongolia, is situated in undulating, sandy rangeland, with an annual rainfall of 360 mm. Owing to over-cultivation and undue collection of firewood, the area of desertification spread rapidly and 81% of the total area had become desertified by the mid-1970s. Mean annual grain yield was only 420 kg/ha. Since the middle of the 1970s, the proportion of land used for dry farming has been regulated, and the area used for plantation and grazing land enlarged. A protective network of plantations, comprising trees, bushes and grass, has been established, including shelter-belts. The area of arable land under threat of sandy desertification was gradually reduced. The plantation work was closely combined with nature conservation. At present, the proportion of land use is regulated as: cropping land 21%, plantation area 27% and pasture 52%. Sandy desertification has been initially controlled. The total yield of food grain has increased by 436%, compared with that before the process of desertification was reversed. Tables 6 and 7 show two more successful examples.

Table 6: Results of Desertification Control at Roledianzi village, Naiman County, Hoeqin Sandy Land, Inner Mongolia

	Before control (1984)	After control (1988)
Shifting sand area	1,000 ha	33.3 ha
Percentage of bush and grass vegetated	10%	30%
Grain yield	150,000 kg	250,000 kg
Average personal income	190 yuan/year	430 yuan/year

Table 7: Results of Desertification Control at Shabianzi village, Yanchi County, Inner Mongolia

	Before control (1984)	After control (1989)
Shifting sand area	1,472.9 ha	1,005.5 ha
Percentage of vegetation on subfixed	7.0%	30.0%
Grass yield on subfixed sandy land	240 kg/ha	1,320 kg/ha
Grain yield	146,000 kg	214,000 kg
Average personal income	399.8 yuan/year	893.8 yuan/year

Another example is Belian village in Fengning County, Hebei Province. The area of the village is 16,870 ha. In 1984 there were 6,800 ha of farmland, 770 ha of forest and 8,930 ha of grassland. In 1987, after the land use pattern had been regulated, there were 4,000 ha of farmland, 1,900 ha of forest and 10,600 ha of grassland. Although the area of farmland decreased by about 41.2%, yields per hectare increased dramatically because of the intensive management of farmlands with better soil and water supply, and the village's total yield was raised by 30%. The 1,300 ha released from cultivation was planted with trees and bushes or grass, the sandy desertification process was controlled, the environment was improved and livestock production developed as well.

Increasing the Rates of Plantation and Pasture

The proportion of land used for plantation and pasture should be increased, with the aim of developing livestock-raising. Both steppe grazing and indoor or semi-indoor feeding systems should be popularised. The latter systems have the benefit of allowing intensive collection of manure. Artificial grassland and forage farms should be created as supplementary sources of fodder to natural grassland. Undulating sandy lands and gentle semi-fixed dunes can be vegetated through conservation and artificial plantation of forage grass; small pieces of fields and pastures can also be used. It is important to increase the linkage between livestock-raising and farming. Straw and manure can be exchanged between these two sectors. Other important steps to control the spread of sandy desertification on rangelands are the determination of carrying capacities of domestic animals, rotation grazing, the reasonable replacement of drinking wells, rearrangement of grazing areas and the construction of paved roads.

Conclusions

To summarise, the measures which can be used to combat desertification are as follows:

Desertification Caused by Wind Erosion

Semi-Arid Zones
 Adjust the structure of land use in dry lands
 Transfer lands to grazing, at a rational carrying capacity

Enlarge forest and pasture and create shelter belts around farmland
Plant trees and shrubs between dunes
Fence off rangeland
Forage farming

Arid zones

Establishment of oasis protection system:
Create protective networks of farmland inside oasis
Make sandbreaks of trees and shrubs on edges of oasis
Establish belt of grass around oasis, protected by fencing
Protection against desert encroachment:
Lay straw checkerboard on dunes and plant shrubs inside it
Build barriers to accumulate sand
Engineering measures against sand encroachment
A plan should be made for the whole drainage area of an inland river, treating it as an environmental unit.

Desertification Caused by Water Erosion

Humid and sub-Humid Zones

Control both gully erosion and slope wash-off by engineering measures combined with biological measures
Stop over-cultivation on severely eroded slopes by fencing and shrub plantation
Establish productive forests such as tea plantations and orchards on slopes

Notes

1 Zhu Zhenda and Wang Tao, 'An Analysis of the Trend of Land Desertification in Northern China During the Last Decade, Based on Examples from Some Typical Areas', *Acta Geographica Sinica*, 45, Science Press, Beijing, 1990, pp. 430–40 (in Chinese).
2 Zhu Zhenda and Liu Shu, *Desertification and its Control in China*, Science Press, Beijing, 1989, 126 pp. (in Chinese).

Chapter 9

Impact of the China–Australia Sheep Research Project, Xinjiang

Improving Sustainability

Frank B. Roseby

The China-Australia Sheep Research Project

This chapter describes a current collaborative project between China and Australia to improve long-term economic benefits of sheep raising in northern China by developing improved and sustainable technologies and facilitating their adoption by livestock producers. New technologies will only be adopted and sustained if they take into consideration the wider interactions between the physical, biological, social, policy and economic environments. To ensure this, the project takes a farming systems approach to the planning and implementation of research, development and extension. Project management must monitor the rapid changes occurring in the economic, policy and social environments to ensure continuing relevance of the project to the needs of livestock producers. The chapter then addresses a problem found in all countries. When the initial funding for a research project ceases, how can we continue through the development and commercialisation stages and ensure the promised impact on the beneficiaries and the economy? The author draws on Australian experience in sustaining research through to commercialisation and adoption of new technologies and encourages research leaders to seek more collaboration with the growing private sector.

Rationale and Location

Wool industry discussions between China and Australia in the early 1980s identified the establishment of a sheep research station in China as an important step towards improving the Chinese wool industry while at the same time assisting communities with some of the lowest incomes in China. The project is located in the Xinjiang Academy of

Animal Science at Urumqi and its associated field station, Nanshan Stud Farm 70 kilometres south-west of Urumqi in the Xinjiang Uighur Autonomous Region of north-west China. On-farm trials, demonstrations and other extension efforts of the project are concentrated in the target area of Changji Prefecture and Urumqi Administration Area. These are representative of the Tien Shan mountains and adjoining grazing and farming areas.

Goal and Objectives

The goal of the project is to improve long-term economic benefits of sheep raising in northern China. To meet this goal the following objectives have been set:

- improvement of genetic material, technologies and management for sheep and wool production by establishing a sheep research centre within the Xinjiang Academy of Animal Science;
- facilitation of the adoption of improvements which benefit sheep raising households in the short and long term by using socio-economic surveys and analysis within the Farming Systems Research and Development approach;
- strengthening the project planning and monitoring capacity of the Xinjiang Academy of Animal Science to support sheep and wool research and extension.

The project should lead to increased production of finewool and meat per head of sheep under situations where the stocking rate has been adjusted to sustainable levels. Many of the new technologies developed will also have flow-on benefits to carpet wool sheep, goats and other ruminants.

Approach

The project was designed to help meet the needs of the *design transfer* phase of technology transfer (Hayami and Ruttan, 1985) to the Chinese sheep industry, hence the emphasis on provision of academic training, laboratories, library and computers. These will strengthen the applied research and extension capacity of the Academy for modification of technologies already proven elsewhere, their validation in this environment and their extension to sheep producers. The farming systems approach taken to research, development and extension ensures that the research and extension elements can be kept in balance and that research is targeted on the real needs of the sheep herders. Close interaction between scientists and farmers is critical and central to this approach, which is described in Byerlee *et al* (1980) and Shaner *et al* (1982). The application of the farming systems approach to research and extension at a site in Central Asia (Gansu Province, P.R. China) is provided by Hardiman and Zhang Xiaohu (1988).

The design of the project recognises the need for advanced research facilities if the Academy is to be at the forefront of sheep and wool research in China, so extensive new laboratories were built and equipped with the most efficient equipment available. However, a strong focus on the social and economic needs of the predominantly Kazakh sheep herders, facilitation of the extension process and monitoring of the impact of new technologies at the household level remain key features of the project.

Current Research Programmes

a) *Breeding*: Finewool sheep breeding research has concentrated on improvement though a programme of genetic selection to increase clean fleece weight facilitated by multiple ovulation, embryo transfer and artificial insemination technologies.

b) *Animal Health*: Roundworm control is necessary to avoid losses and research has developed a method of treatment which is appropriate to Xinjiang conditions. Improvements in lambing management are expected to bring immediate benefits to the herders by improving lamb survival and life time productivity through better peri-natal health and nutrition. A longer-term objective of the Health Programme is to contribute to the development of a vaccine for sheep or dogs to control hydatids. The incidence is very high in Xinjiang, and although not a problem for the sheep, the disease can be fatal to humans.

c) *Nutrition*: The objectives of the Nutrition Programme are to measure seasonal variation in pasture production and its relation-ship to animal productivity; to improve the utilisation of low quality autumn forage with high protein feed supplements; and to estimate the balance between nutrient intake from mid-winter grazing and nutrient requirements for thermo-regulation and grazing activity.

d) *Pasture*: Initial reports by the rangeland and pasture specialists discussed the apparent deterioration occurring in Xinjiang pastures and highlighted the importance to the herder communities of establishing sustainable grazing management systems. The aim of the pasture research programme is to increase pasture production and help develop these sustainable grazing management systems. Specific activities will identify grass species that maintain their herbage quality as standing hay for winter grazing; use a computer model to relate livestock productivity to available forage resources; determine the optimum level of forage utilisation on summer grazing lands as a basis for estimating sustainable stocking rates; and improve weed control.

e) *Reproduction*: The aim of the reproduction research is to increase the number of lambs weaned per ewe. Specific research activities address relationships between ewe mating body weight and net productive rate; seasonal variation in oestrous and ovulation; and twinning rate and lamb survival. The real risk that this technology could lead to higher sheep populations is being addressed by the project.

f) *Wool Technology*: Wool technology research is improving our understanding of the relationships between wool quality and the environment. The wool laboratory being established in the Academy will provide objective wool quality assessment for producers, buyers and processors of wool. Currently, any value added to wool by herders can be lost because the market is not efficient and does not always reward better quality with a higher price. As the efficiency of the market improves, the benefits of producing better prepared wool will remain with the herder.

g) *Economics*: The Economics research programme is ensuring a better knowledge and understanding of the economic environment in which the herders operate and is helping to ensure allocation of research resources to those problems of most concern to the herders.

Continuation Programmes

The achievement of the long term goal of the project – improved economic benefits for sheep raising in northern China – will be only brought about if technologies developed by the project are adopted by the sheep producers, initially in the project area and eventually throughout northern China. Adoption of improvements will take place if two main outputs are achieved. Firstly, there is the need to understand the problems and needs of herding households at the 'grass roots' level through surveys to determine their constraints, and studies to understand their response to adoption of new technologies. Secondly, there is a need to undertake a process of education and dissemination of information which encourages adoption of new technologies by herders. That is, the process known as *extension*. This involves the trials using these new technologies at the farm level by research staff and their modification to meet local needs. As herding households begin using the new technologies they will need guidance from trained extension workers. Thus, in the second half of the project there is much greater emphasis on training of research and extension staff in the best way to make and use:

- audio-visual material to promote awareness of new technologies;
- on-farm trials to ensure they are commercially sustainable;
- demonstrations of their suitability to the herders' own situation.

Impact of Training on Research

Training of Chinese staff in Australia on many aspects of the sheep industry is vital to the development of the Academy as a leading sheep and wool research and extension centre. Three researchers are enrolled in Masters Degrees and five are enrolled in Post-graduate Diplomas. Another twenty-six research, extension and management staff of the project are spending from three to six months working in Australian sheep research and extension teams. Thirteen senior managers from the Academy and associated institutions have participated in study tours to become more familiar with the Australian sheep industry.

Although the project is now virtually fully staffed, the large number of key research staff currently receiving post-graduate training in Australia is limiting research progress. This was entirely predictable and is a direct consequence of the project design. It is always difficult to balance the needs for staff training with the need for the same staff to be active on the project.

Impact of Economic Reform on Extension

China's rapid economic reform (Du Runsheng, 1989) is giving herders much more independence in their work and making them more responsible for any costs involved. They will adopt new technologies only if they are sure they will benefit, so we now need to demonstrate clearly these benefits in the herders' own locations. In essence, extension staff can no longer tell farmers and herders what to do or do it for them. The situation in Xinjiang is now very like the Australian situations where farmers need to be *encouraged to adopt* new technologies and pay any costs themselves. As extension staff come to terms with this impact of economic reform, project emphasis in extension has shifted from input supply to using education and training to change behaviour and attitudes. Australian expertise is being used to demonstrate preparation of training materials (pamphlets and video films) and use of on-farm trials and demonstrations for extension. Trainers are being identified within the predominantly Kazakh communities at extension sites and taught effective educational methods. Adoption is more likely if the message is delivered in a friendly style, by a person with credibility, in the most familiar language.

Impact of the project on Women

From the outset, women's position within the family and society has been studied with a view to assessing the impact of the project on women and to ensure that women's roles and needs are understood and addressed. These studies have identified the following essential elements: the long-term need to improve women's educational levels, their need for training and their need for additional employment opportunities. A simple framework has been established so that the Economics Unit can monitor the growing impact of the project on women's technical role, economic status and social status. Addressing some of these identified needs of the herder women, a women's training specialist recently developed and demonstrated a training course in fleece preparation, a traditional role of the herder women. Sustainability of the pastoral systems depends on a strong commitment from and participation by women. To ensure this the status of women must be enhanced and women must continue to participate in decision-making at all levels.

Impact of the project on the Pastoral Environment

Current grazing management systems in the pastoral regions of China are not considered sustainable and deterioration of the pasture is apparent (Longworth and Williamson, 1993). The early introduction of stable systems of grazing management where stocking rate is matched to pasture production is imperative. However, systems which depend solely on regulations and the force of law to ensure their adoption will not succeed in the long term. Appropriate economic incentives must also be in place and the herders must feel confident they will stay in place. Then we may expect to see smaller household flocks of higher-producing sheep with similar wool and meat production per flock, but of higher quality and value.

In attempting to improve grazing management, research activities often focus on a particular component of production and may not pay sufficient attention to the impact the new technology may have on the biological, physical or social environment. Because this impact could be damaging to the fragile pastoral environment, a thorough assessment of the environmental impact of all research activities is necessary. Ideally this should be done prior to their commencement. However in reality the impact may not be immediately apparent so a periodic project wide environmental impact assessment has proved to be more practical. In the long term, the network of pasture monitoring sites being established in the project target area will provide the empirical data which will enable us to determine the sustainability of the existing grazing management systems and those introduced.

Funding of Research, Development and Extension

Traditionally, funding of all research by governments has been provided to academic and government research institutions which have decided what research they shall do using traditional methods of priority setting (e.g. see Davis and Ryan, 1987), guided by public policy. Continued funding was usually dependent on research performance judged by other researchers and administrators.

Few governments have been prepared to put significant funding into development and commercialisation of research results applicable to manufacturing, leaving it to individual firms to decide which results, if any, should be taken further. Most governments view agriculture as a special case and provide some funding for trialing and extension of new technologies, but this funding is always limited and in recent years has declined dramatically in many Western countries (Watson *et al*, 1992).

Research alone brings little direct benefit to society. However, its development and application to productive enterprises in agriculture and industry can bring enormous benefits. Such application is most likely when linkages between the research sector and the productive sectors are strong and the productive sectors can exert a strong influence on the type of research done. Unfortunately, in Australia, and to an even greater extent in China (Conroy, 1992), these linkages are not strong. In both countries the result has been the same: a lot of excellent research has been done but often the results have not been developed and applied to industry or extended to farmers.

Australian Experience in Industry and Agriculture

To help overcome this problem of weak linkages between research institutions and firms or farmers who could use their research results, the Australian government has made more of its funding of research institutions dependent on them receiving a matching contribution from potential beneficiaries. The long-term aim is to stimulate research by those best able to use the results. However, firms and farmers usually do not have appropriate facilities to carry out research, so funds contributed by firms and groups of farmers are matched with funds from government and provided to institutions with research capability prepared to do the work for them.

In Australia, those institutions collaborating with firms and farmers to do their research are growing, while those unable to respond to the needs of the productive sector are disappearing. Making a research institution's funding dependent on its close collaboration with its clients has of course assisted governments with the funding of research. In addition, it has ensured that the research done meets the needs of the clients in productive sectors who can use the results to develop new technologies and generate benefits for themselves and the community. The other major benefit is to ensure better accountability of public funds spent on research.

In Central Asia, although it may be too early to expect a financial contribution from farmers, they can still be involved in deciding what research is done. Research institutions should adopt a farming systems approach to the planning of their research to ensure that the needs of the intended beneficiaries are met. Other potential beneficiaries of industrial and agricultural research such as suppliers of fertiliser, agricultural chemicals, equipment and feedstuffs should be encouraged to collaborate with research institutions for their mutual benefit and long-term benefits to the people.

References

Byerlee, D., Collinson M. *et al* (1980), *Planning Technologies Appropriate to Farmers: Concepts and Procedures*, CIMMYT, Mexico.

Conroy, R. (1992), *Technological Change in China*. OECD, Paris.

Davis, J.S. and Ryan, J.G. (1989), 'Evaluation of and Priority Setting for Agricultural Research: Methodology and a Preliminary Application for China' in Longworth, J.W. (ed.), *China's Rural Development Miracle*, University of Queensland Press, St. Lucia.

Du Runsheng (1989), 'Advancing Amidst Reform' in Longworth, J.W. (ed.), *China's Rural Development Miracle*, University of Queensland Press, St. Lucia.

Hardiman, R.T. and Zhang Xiaohu (1988), *Farming Systems Research and Extension in Central North China*, Proceedings of the International Workshop on Farming Systems Research and Development, USA, 1988.

Hayami, Y. and Ruttan, V.W. (1985), *Agricultural Development: An International Perspective*, The Johns Hopkins Press, Baltimore, USA.

Longworth, J.W. and Williamson, G.J. (1993), *China's Pastoral Region*, CAB International and ACIAR, Canberra.

Shaner, W.W., Philipp, P.F. and Schmehl, W.R. (1982), *Farming Systems Research and Development: Guidelines for Developing Countries*, Westview Press, Boulder, Colorado, USA.

Watson, A., Hely, R. *et al* (1992), *Review of Field Based Services in the Victorian Department of Food and Agriculture*, Agmedia, Melbourne, Australia.

Chapter 10

Environmental Problems in Kazakhstan

Aliya S. Beisenova

Overview of the Environment in Kazakhstan

The current state of the environment in Kazakhstan is extremely unsafe. For a long time, the territory of our Republic was considered huge and wild; it was thought perfectly acceptable to situate nuclear test-sites and space-vehicle launching grounds here and to develop various branches of industry without taking the environment into account. Even very recently, we were unaware of anything except impressive figures indicating the 'industrial might' of the Republic, and were sometimes told that the panorama of smoking chimneys was 'a joyous sight'. We completely forgot that statistics for ill-health were rising. None of the regions of Kazakhstan where these activities occur has escaped acute environmental problems.

In the Karaganda industrial area, slag heaps take up enormous areas of fertile land. In the waste dumps, coal smoulders slowly, emitting 250,000 tonnes of carbon dioxide and smoke into the atmosphere every year. This region is one of the worst in the former USSR for cardio-vascular illnesses, high blood pressure and lung disease.

In Aktyubinsk the health indicators are as poor as those in Karaganda. Over 50% of the polluting substances are produced by the city's chemical plants, which make ferrous alloys and oil products. Dust, gas and heavy metals drip into the soil, plants and animals, and most seriously, people suffer.

The natural environment of the Caspian region is under considerable pressure from human activity. The concentrations of hydrogen sulphide and sulphurous gases are several times higher than the acceptable level, especially near the Tengiz oilfield. The pollution of the northern shore of the Caspian Sea and its tributaries, the rivers Ilek, Emba and Ural, with oil products has resulted in mass deaths of waterfowl, illness in sea animals, and a sharp decrease in the numbers of sturgeon. It is also reflected in the health of people living in the region.

During experiments at the nuclear testing site near Semipalatinsk, the level of radiation is increased, which affects the health of humans and animals, and the condition of the topsoil. Every time a spacecraft is launched at the Baikonur cosmodrome, a huge quantity of harmful substances is emitted into the atmosphere, as a consequence of which a wide area of land is left unsuitable even for livestock.

One of the most important regions for phosphorite mining in the world is the Karatau-Zhambyl industrial area. Zhambyl has a number of plants producing phosphate and superphosphate fertilisers, which emit sulphur-bearing gases and carbon dioxide into the atmosphere. The wind carries dust from the phosphate quarries over enormous distances, and the exhausted quarries are left as open sores on the body of the Earth.

The situation in Shymkent is no better. Standards in a number of industrial enterprises do not conform to the requirements of legislation on nature conservation.

In the south and south-east of Kazakhstan, there are a number of non-ferrous metallurgy plants. The copper-smelting factories of Dzhezkazgan, Balkhash and Ust-Kamenogorsk pollute the atmosphere with emissions of sulphur, carbon oxides and particulates. The existing cleaning equipment is out of date and does not work at full capacity. The Balkhash factory dumps some of its waste into Lake Balkhash, where fish are poisoned and die. This is also endangering human health.

All the industrial processes which take place on the river Ili and its tributaries automatically tell on the ecology of Lake Balkhash, Kazakhstan's largest lake. Balkhash supplies 10,000 tonnes of high-quality fish products a year, that is more than a third of all the fish caught in the waters of the Republic. The problem of how to preserve the Lake has a peculiar ecological and social significance in Kazakhstan today. People are asking what will become of Lake Balkhash if the volume of water consumption remains as it has been, if the waters of Balkhash continue to be polluted with deadly poisons washed from the fields, if the copper-smelting plant in the city of Balkhash continues to dump its effluent into the Lake, if the plan to reconstruct the delta of the river Ili goes ahead? If all these outrages are not stopped, they could lead to the transformation of the western part of the lake into an industrial reservoir, and its eastern part into saline dust, and finally to the total loss of the Lake's value to the fishing industry. Already today, fish from Balkhash is dangerous to eat. But it is officially forbidden to say so. We cannot be silent, the disaster of the Aral Sea must not be repeated.

The environmental problems of the Balkhash basin are directly linked to the ecology of the topsoil. It is no secret that agriculture in Kazakhstan is 'developing' extensively. The harvest is increased by ploughing up virgin and fallow lands. As a consequence of this, the diversity and quantity of plants is reduced.

The Almaty *oblast* (province) straddles three types of vegetation: desert, semi-desert and steppe. Thus, if its lands are to be cultivated they must be irrigated. However, it is impossible not to be aware that the expanses of irrigated land have not justified all the hopes they promised at first: water has been aimlessly squandered and millions upon millions of the nation's money have essentially been buried in the earth. Throughout these sandy expanses huge numbers of river fish are dying. In Chingildy the soil is becoming so badly salinated that in a matter of decades the land could become unfit for agriculture; a similar fate could await Malaysardy. The restoration of these lands in the future will require more and more money.

Another acute ecological problem is the immoderate use of poisonous chemicals on the fields in the province. DDT and hexachlorane, both highly toxic and dangerous to human health, are in daily agricultural use, damaging the genetic structure, causing dangerous chronic diseases such as cancer, hepatitis, bronchitis, cyrrhosis of the liver, heart disease and mental illness. All this is true, but nevertheless, farms continue to use hundreds of tonnes of these chemicals every year. The use of sewage from sedimentation tanks for irrigating the land has created a new 'laboratory', for testing the effect of harmful substances on human health in a chain: human sewage – fields – feed – cattle – humans.

Scientists affirm that the condition of the environment is now such that the probability of dying from it is practically the same as the probability of dying on the roads. The risk is exceeding the limit of acceptability: one in 10,000. But, whereas we do not have to fly in aeroplanes or travel in cars, we cannot avoid breathing and eating. Thus, an ecological approach to the solution of all questions is simply unavoidable today.

Water Supply

Kazakhstan's water system gives cause for particular concern. The old Aral Sea is no more: its surface area has been reduced by nearly half.

In the last few years all the water from the Amu Darya and the Syr Darya, its principal tributaries, has been used for irrigation. It has been consumed beyond all measure. In places, the water has already receded from its former shores by 100 kilometres. In its wake, the subsoil waters have been salinised and are likewise receding, and plant and animal life is disappearing. From the exposed seabed, the wind picks up tens of millions of tonnes of caustic salts and scatters them on the ground for distances up to 400–500 km from the Sea. These salt-dust clouds are threatening to destroy the fields for whose sake the water was collected in the first place. Salt from the Aral Sea has been detected as far away as the glaciers of the Pamirs and Tien-Shan, where it accelerates their melting. The drying up of the Aral Sea is not only an ecological disaster for the peoples of Kazakhstan and Central Asia; it is a catastrophe on a global scale.

Mention should be made of the water quality in the Syr Darya. Drinking water is taken from it to this day, though the water has long been contaminated, 35% to 40% of the waste is industrial effluents, city sewage and drainage waters. The river permanently contains agents which cause dysentery, typhoid fever, salmonellosis and hepatitis. There are also dangerous chemicals in the Syr Darya, such as DDT and hexachlorane. Fish from the Aral region should not be used as food.

Kazakhstan's large rivers – the Irtysh, Ili and Ishim – have become drastically shallow because of the water taken for irrigation; but this heavy water use is not always justified. Used water from the fields goes back into the rivers, heavily polluted with pesticides and fertilisers. Using this water for the needs of the population is dangerous. The fish are contaminated and their diversity is impoverished.

All branches of industry, as well as agriculture, manufacture their products to improve people's material well-being, but in the end such poor treatment of nature turns into a terrible disaster for people and for nature as a whole.

Almaty

The former capital city is no exception to this pattern of environmental degradation. Emissions of harmful substances into Almaty's atmosphere from transport and stationary sources come to 303,800 tonnes a year, including 41,000 tonnes of solids: Almaty is surrounded by mountains, and ventilated only by mountain-valley winds, in other words, hardly at all. For this reason the problem of the pollution of the atmosphere and surrounding land with industrial wastes has deepened as the city has developed. Two basic 'historical mistakes' were made in the planning of the city: an incorrect choice of site and the decision to locate heavy industry in Almaty.

In the nineteenth-century, settlers were attracted to what was initially called Vernoje by its environment: the abundance of sunny days, good land and mountain rivers, which were important for agriculture. Right up to the 1917 Russian Revolution, the most important occupation of the city's inhabitants, after government service and trade, was agriculture: market gardening and livestock-raising. Was this ill-ventilated foothill valley the right choice of site for a city? Now the meteorologists say: 'Of course not!' But it would be wrong to lay the blame entirely on Tsarist functionaries, because this first 'historical mistake' was not the last to contribute to today's complex of problems. Smaller mistakes followed, one after another. The main one was the 'Leningrad Project' for building up the centre of the city. This project had worked well in Leningrad, where there are frequent winds from the Gulf of Finland, but in Almaty it was carried out without consideration of the circulation of the city's atmosphere.

For twenty years Almaty's sky was blackened by Brick Factory Number 3, and for a long time, by the rubbish recycling plant, ADK; a combined heat and power (CHP) station was built, in violation of health standards, in a residential part of town; planting with trees and gardens was carried out in a haphazard and ignorant fashion. Imported ethylised petrol has been used for a long time; according to calculations by Swiss scientists, every litre of this petrol burned in a car releases 274–405 mg of lead bromide, which is harmful to human health. It is enough to say that in the capital every 24 hours, for every inhabitant, transport emits up to half a kilogramme of pollutants into the atmosphere; in other words, the city's people usually breathe air which has come through the exhaust pipes of cars and factories.

According to data prepared by the Department for Control of the Use and Conservation of Atmospheric Air of the Kazakh SSR's State Committee for Nature, transport alone emitted 164,800 tonnes of harmful substances into the capital's atmosphere annually, and CHP Stations nos. 1 and 2 emitted over 4,000 tonnes of harmful substances a year. There are several major industrial enterprises in the city. The tobacco processing plant is situated in the centre of the city. There are forty functioning cleaning installations, of which fourteen are in a state of disrepair. At the Asphaltobeton concrete factory, only 77.1% of harmful substances were removed from the flue gases. The Kirov car factory is also situated in the centre of the city: only 42% of its sources of harmful substances were equipped with cleaning devices. The furniture company Alma-Ata is releasing 500 tonnes of harmful substances a year into the atmosphere.

It is perfectly clear from these statistics that the most polluted areas of the city can be identified even without using instruments. These are: the districts near the CHP Stations, the airport, the railway stations (Almaty 1 and 50-let Oktyabrya), also the streets where the private sector predominates and where traffic is most intense. For a long time, Prospekt Abaya was an exception in this respect, but now nothing differentiates it from the other busy thoroughfares of the city.

As a result of the fact that the city itself is very unsafe in environmental terms, it also pollutes the surrounding area. Thus, for instance, dirty air leaves the city along Prospekt Lenina and reaches the mountains, where it melts the snow and ice and pushes back the snow-line. Intensive melting of snow creates a danger of seasonal mountain torrents, and, due to the unstable tectonic situation (the Maloye Alma-Atinskoye ravine is situated on a tectonic cleft) this danger is increased many times. The action of dirty air on the snow-line drives the whole spectrum of ecological zones up the mountain-sides. The boundaries of each zone of vegetation are getting higher, with the result that there is a danger of the fir tree disappearing; and it is well known that one hectare of pine forest can clean 18 million m^3 of air a year by trapping 400 tonnes of dust.

It can only be expected that problems of water-supply will become very acute in Almaty. Attention should be paid to how we use water for our everyday needs. This applies both to cold and to hot water. To provide hot water for the city, a colossal amount of energy is consumed, in the form of a huge quantity of coal, the burning of which poisons the air basin of the city.

Some time ago, the Zhamankum storage-lake burst its banks. This reservoir is the main settling-tank for Lake Sorbulak. For almost twenty years, contaminated water had been gathering there; all of this water was washed into the River Kaskelen and the reservoir. The Kaskelen is one of the tributaries of the river Ili: sanitary standards were thus inevitably very low. Some of the contaminating substances which had gathered in the sands of Lake Zhamankum were also washed into the Kaskelen and into the reservoir, and will continue to contaminate the water of the Kapchagay for a long time to come. This will undoubtedly also be reflected in Lake Balkhash. The situation in the Sorbulak and Zhamankum region has still not been stabilised. The threat of Lake Sorbulak bursting its banks is still hanging over us: about 1 million m^3 of sewage has gathered in it, and if this problem is not solved urgently there is a serious risk of a disaster. Until the breech in Lake Zhamankum is blocked, sewage sediments will continue to be washed into the Kaskelen.

Nature has the capacity to clean itself naturally; this self-cleansing continues throughout the year, but is most intensive in the warmer part of the year. Unfortunately, the quantity of industrial waste generated exceeds the possibilities of natural cleaning. A vital part is played in the self-cleaning process by vegetation. It has been calculated that 25 m^2 of planted area is needed per inhabitant to restore the oxygen balance of a large city. But the 'green air filter' in the city only works in the summer; in winter the city's inhabitants undergo latent oxygen deprivation.

The mountain-valley circulation of air also plays a role. These currents are strongest in summer, in clear weather, when mountain breezes prevail in the weakly-heated valleys, open to the north and northeast. In winter, the difference between the temperature of the mountains and the valleys is less pronounced, so that the speed and power of the mountain-valley convection currents are reduced, and the built-up area of the city is less well ventilated.

Unfortunately, planners, architects and power engineering specialists have taken advantage of the atmosphere's ability to clean itself to plan 'organised emission' installations. The construction of the city up to 1950 was conducted with the mountain-valley currents and the seismic situation in mind. In recent years, however, high rise buildings have appeared which block the paths of the mountain winds, such as the residential blocks on Vinogradov Street and Kommunistichesky, Kirov and Mir Prospekts, the Pushkin Library, the Lermontov Theatre, the Central Committee Building, the television studio and so on. Environmentally ignorant building development is one of the most important causes of pollution in the city. The most glaring example of this is the construction of the Samal *mikroraion* (housing estate), which blocks the only route for mountain-valley air circulation in the eastern part of town, the ravine of the Malaya Alma-Atinskaya River. Apple orchards are being barbarously destroyed, Tien Shan fir trees and Kolpakov poplars planted a hundred years ago are perishing, and in their place concrete buildings are being put up.

Conclusions

To sum up, Almaty's environmental problems are becoming ever more acute, and this process is encouraged by the narrow departmental approach to problems. The following are some suggestions for improving the environmental situation in Almaty:

- accelerated construction of the metro
- further introduction of gas heating systems
- increasing the capacity of the public transport system
- planting new woods, hedges and flower beds
- a ban on bringing ethylised petrol into the capital
- economising on fuel by reducing automobile trips
- widespread demolition of residential blocks
- building new suburbs, taking into account the circulation of the lowest layers of air
- preservation of the ravines of the Bolshaya and Malaya Alma-Atinskaya Rivers as areas of recreation, and the cessation of building development in them.

At the present time, our society needs, more than anything else, reliable information about the ecological situation in every locality and region. There is now a wide debate about the problems of the Aral Sea and Lake Balkhash, generating a great deal of senseless noise. But people's attention is fixed only on these regions. What is being said on our streets and in our cities is even more important. Information on pollution throughout the Republic should be made openly available to the whole population. Then public awareness will be more productive and the authorities will do more to solve the problems.

The Abay Pedagogical Institute, as one of the Republic's leading institutions of higher education, should lead the work of environmental education for future teachers. Today it is essential to expand the Laboratory for the Conservation of Nature and the Ecology of Landscapes in the Department of Physical Geography and orient it towards problem-solving.

Chapter 11

Sustainable Mineral-Driven Development in Turkmenistan

Richard M. Auty

This chapter examines the development potential of Turkmenistan in the light of its natural resource endowment and social capital. It proceeds by summarising the natural resource endowment and then evaluates the post-Soviet development strategy, noting the consequences of an unexpected deterioration in gas markets. Finally, a revised development strategy is outlined in the context of the on-going debate concerning the desirable pace of reform in the transition economies. The chapter assumes, in line with neo-liberal economic theory, that the optimal development strategy requires Turkmenistan to realise its comparative advantage as an exporter of a range of primary products to both the western Asian region (notably Turkey) and also to adjacent global markets in Western Europe, South Asia and East Asia.

The Natural Resource Endowment

Classifications of countries by their natural resource endowment have used different criteria, including: land area (Wood and Berge 1994); export orientation and population size (Syrquin and Chenery 1989); and per capita cropland and domestic market size (Auty 1994). This chapter opts for domestic market size and per capita cropland because these two factors, suitably adjusted to accommodate variations in per capita income and mineral resources, respectively, define the critical constraints on both manufacturing and export opportunities. By these criteria, Turkmenistan is a small, resource-rich country. It has deficiencies in social capital arising from the legacy of the restrictions on the effective trading and exchange of goods under the Soviet system.

More specifically, Turkmenistan is a landlocked country which, despite its large geographical area of 488,000 km^2, is only moderately land-rich, with 0.38 hectares of cropland per capita (WRI 1996). Cropland extends over barely 2% of the land surface and meadows and pasture occupy 69% of the land surface. However, the country has large reserves of hydrocarbons, especially natural gas. Gas processing offers a route to industrialization which is in addition to those of import substitution and crop processing for export. However, the small domestic population of barely 4.5 million with its US$500 per capita income affords a very small market for import substitution, while the landlocked location reinforces distance from major markets to weaken prospects for industrial exports. Consequently, the production and export of primary products is likely to form the basis of the country's economic development through the next generation at least. It should be noted, however, that this development strategy has been highly successful in countries as culturally varied as New Zealand, Australia, Canada, Chile and Botswana.

The economic links with the Soviet Union extracted cotton and hydrocarbons from Turkmenistan in exchange for manufactured goods. These links left agriculture responsible for almost 50% of GDP and two-fifths of employment in the early-1990s (US Department of State 1994). Somewhat surprisingly, however, the cropped area had contracted through the 1980s, so that at independence, the total area under cultivation was 1.48 million hectares compared to 2.1 million hectares in 1983 (WRI 1996). This decline in cropland was accompanied by the concentration of crop production on irrigated land whose share of the total cropped land consequently rose from 48% to 91% over that same period.

Like many low-income countries, agricultural production in Turkmenistan is relatively undiversified and heavily dependent on a single product, in this case, cotton. In the early-1990s, 1.3 million tonnes of cotton were produced from 45% of the arable land: it generated 56% of farm output and 16% of total exports. Wheat, the next most important crop occupied one-eighth of the cropped area in 1993 and yielded 374,000 tonnes. However, yields were low, and the collective farms were highly mechanised, despite the low cost of rural labour. For example, the wheat yield, although twice that of Kazakhstan to the north, averaged only 2 tonnes per hectare which was barely two-fifths of the industrial country levels (WRI 1996). Domestic grain production needed to be supplemented by imports, in part because animal feed absorbed two-fifths of Turkmenistan's grain demand (WRI 1996). Livestock accounted for 22% of agricultural output, but less than one-twelfth of exports, and was based upon a cattle herd estimated at 960,000 1992–94 and a combined sheep and goat herd of 6 million animals.

Despite its dominance of production and exports, the farm sector generated only 23% of total exports, of which cotton alone accounted for two-thirds. This is because natural gas production and processing earned 70% of the country's foreign exchange in 1993 and oil exports accounted for an additional 7% of exports. Natural gas also generated 70% of Turkmenistan's industrial output. Recent estimates by BP (1996) suggest the country's natural gas reserves are 101 trillion ft^3 (which is 2.9 trillion m^3 and equivalent to 2.6 billion tonnes of oil – or 19 billion barrels). The reserves comprise 2% of total global gas reserves and are the eleventh largest in the world. However, if *possible* reserves are also considered, then the total natural gas reserves may be five times this level. As it is, the proven reserves alone can support over 30 years of production at the 1993 rate of extraction of 65 billion m^3 (22 million tonnes of oil equivalent (IMF 1997)).

The oil reserves of Turkmenistan are believed to be much smaller than the natural gas reserves: the proven oil reserves are 190 million tonnes, or 1.4 billion barrels. However, exploration and production were both neglected after the 1960s. Annual oil production fell from 300,000 barrels per day (bpd) in the 1970s to 85,000 bpd in the early-1990s. In 1993, domestic consumption was 4.3 million m^3. Refined products were produced in two unsophisticated oil refineries, each with a capacity of around 120,000 bpd, with some exports by rail and small tankers. Other minerals, in addition to natural gas and oil resources, include 800 million tonnes of coal reserves which could substitute for annual coal imports that were 110,000 tonnes in the mid-1990s. There are also commercial reserves of precious metals, including silver, gold and platinum.

Faced with this resource endowment and an economy and society in need of thorough-going reform, the newly independent Turkmenistan government opted for a strategy of gas-driven development in 1993. The government hoped that expanded natural gas production together with diversification into new markets would generate sufficient additional revenues that investment could be boosted to sustain rapidly-rising incomes, while the extra foreign exchange from gas exports would ease the import constraint on economic development. It was expected that the additional rents from natural gas would allow the country to reduce its heavy dependence on imports and permit a high degree of economic autonomy in regard to both food and industrial production. In addition, the natural gas rents would permit the pace of social change to proceed slowly: a 'big bang' reform was ruled out in favour of a gradualistic approach.

The Gas-based Development Strategy 1993–97

The Initial Hydrocarbon Export Plan

The Turkmenistan government launched an ambitious ten-year development plan in 1993 which was designed to achieve the country's potential as a 'second Kuwait' as quickly as possible. The plan was expected to double per capita income to almost US$6000 in purchasing power parity terms by 2002 (*Financial Times* 1997). It was launched upon a base of healthy financial reserves and good prospects for expanding gas capacity and markets. The back-bone of the development plan called for natural gas production to reach 130 billion m^3 by the year 2000, and 230 billion by 2010 (Sagers 1993). It also aimed to double oil production by 1996 (RPI 1996), and thereafter to lift it to 560,000 bpd (204 million barrels per year) by the year 2000, and to 1.5 million bpd by 2010 (Sagers 1994). It should be noted, however, that these ambitious plans are based on estimated oil reserves of 46 billion barrels of oil, a figure considered to be highly optimistic by external observers (Sagers 1994). Finally, the development plan also called for cotton production to expand by one-third, to 2 million tonnes, and for greater industrial production through boosting added value on the country's primary products, notably cotton.

Initial priority in the plan was given to the attraction of foreign investment to expand gas production, rather than to improvements in social welfare. The first stage of the development plan called for an investment of US$5.5 billion to be made by 1996. The government planners projected that such a level of investment would earn profits of US$2.2 billion by the year 2000 and thereby become self-financing (Sagers 1994). In order to attract investment in gas production, attention was given to the basic infrastructure. A national airline was set up and a new airport was built in the mid-1990s along with new roads, hotels and government buildings in the capital city. Once these projects were well under way, the emphasis shifted to the construction of new pipelines in order to avoid dependence upon the Russian gas network and to diversify export markets.

The Transportation Constraint on Natural Gas Rents

Despite considerable set-backs, the Turkmen government still believed in 1997 that by the year 2000 it was technically possible to export up to 100 billion m^3 of natural gas to the traditional CIS markets and an additional 30 billion m^3 of gas from fields in eastern Turkmenistan to South Asia, and a similar volume from the western part of the country to Iran and Turkey. This would lift total gas production towards 160 billion m^3. The potential rents this would engender can be calculated by deducting from the delivered price of gas, the cost of extraction and the cost of transport. The cost of gas extraction in Turkmenistan are estimated by the World Bank to be 50–55^c/thousand feet^3 (mcf), or some US\$16–20 per thousand cubic metres (mm^3). Delivered prices of natural gas range from US\$2.25/mcf (around US\$80/mm^3) in the western FSU to US\$2.50–3.00/mcf (US\$89–107/mm^3) on world markets.

The above figures imply that transport costs must not exceed US\$1.75/mcf (US\$62/mm^3) on CIS sales and US\$2.50/mcf (US\$89/mm^3) on world markets in order to generate even the smallest rent. A rule-of-thumb transport cost through the existing FSU network is 40^c/mcf per 1000 kms (US\$15 per mm^3). Gas exports into Russia would incur total costs of US\$1.80/mcf using the existing network, or US\$2.05/mcf for a new pipeline. These prices would generate from 45^c/mcf to 20^c/mcf in rent. However, the Russian gas monopoly, Gazprom, shut Turkmenistan out of routes into southern Russia in 1993, despite the Turkmenistan claim to own 12% of the shared pipeline network. Worse, payments delays emerged within the reforming economies elsewhere within the CIS. These unforeseen developments led to the collapse of traditional gas export markets through the mid-1990s and intensified the need to build new pipeline capacity.

Five natural gas pipeline routes were actively considered (Table 1), but their viability and their capacity to generate rents, varies. The first new pipeline to be targeted was a modest 150 km link into Iran. Sales to Iran might reach 2.5 billion m^3 and, by displacing domestic Iranian fuel oil consumption, free that oil for export. This is very advantageous for Iran because liquid hydro-carbons have significantly lower transportation costs per unit heat value than natural gas. However, Turkmenistan can reap greater benefits from a subsequent agreement, signed in 1997, to extend the Iranian gas pipeline westwards into Turkey. This line would initially add an extra 8

Table 1: Turkmenistan's Proposed Oil and Gas Pipelines

Destination	Route	Length (kms)	Size (b m³)	Estimated Cost ($b)
Turkey	Iran	1400	15–28	3.7–5.0
Turkey	Azerbaijan, Georgia	1300	2–7	1.5[a]
Pakistan	Afghanistan	1500	20	2.5–3.0
Japan	China	6100	28	12.0–16.0
Europe	Iran, Turkey	4084	n.a.	11.0[a]

Sources: IMF (1997), except a: Sagers (1994), 52

billion m^3 in volume.[1] However, the additional distance would push total transport costs to US$2.25/mcf, leaving only a modest rent of 25c/mcf or less.

More long-term, the western pipeline can be extended into Germany and, although the delivered cost would rise above US$3/mcf, that figure might be reduced in two ways. First, the extended pipeline might be financed by export credits bearing only 7–8% interest rather than the private return on capital of 15%, a key consideration for a project whose costs are dominated by capital charges. Second, a contract for large sales volumes would capture the economies of scale. Unit rents would still remain minimal, however, in the absence of real increases in long-term energy prices (a development considered unlikely (World Bank 1997)).

Potentially higher rents could be secured from a 1600 km pipeline from the large Dauletabad field through Afghanistan to link into the Sui gas field in Baluchistan and on to Karachi (Table 1). This pipeline is being actively investigated by Unocal and the governments of Pakistan and Afghanistan. In this case, the economics of transportation are feasible, but it may not be possible to secure the financing, given political instability in Afghanistan. Meanwhile, a proposed eastern pipeline through China, does not appear to be viable because it cannot compete with either LNG shipped from southeast Asia or natural gas piped from Eastern Russia.

These gas transportation costs illustrate an important weakness in the gas-based development strategy which arises out of a critical distinction between the cost of transporting oil and the cost of transporting natural gas. The cost of transporting natural gas is much higher, per unit of heat value, than the cost of transporting oil (Stauffer 1975). It follows that the potential rents per unit heat value on natural gas are lower than the potential rents on oil. More specifically, the costs of transporting natural gas by pipeline from Turkmenistan over the most favourable routes seem likely to absorb at least two-thirds of the total revenue. The costs of gas extraction will absorb a further one-sixth, leaving no more than one-sixth of the revenue as rent. In contrast, the cost of oil transportation is unlikely to exceed one-fifth of the revenue while extraction costs under typical Middle Eastern conditions would absorb less than one-fifth. This leaves more than three-fifths of the oil revenue as rent. Basically, the potential rent per unit of natural gas is likely to be less than one-fifth that of oil *under the most favourable of gas transport routes* open to Turkmenistan.

The Scale of the Potential Natural Gas Rents

Nevertheless, given the small size of the Turkmen population and the low level of per capita income, the expansion of natural gas production could still generate revenues on a scale sufficient to strongly impact the country's early stages of development. Table 2 estimates the potential rent stream from natural gas which might have been secured by the late-1990s had the gas-based development strategy encountered fewer obstacles. First, the alignment of domestic energy prices with international levels would generate around US$800 million in rents on the 10 billion m^3 consumed within Turkmenistan. But, more importantly, it would also provide strong incentives towards the more efficient use of energy within the domestic economy. In addition, the restoration of pre-reform sales into the neighbouring CIS countries at international prices would generate rents around US$1.6 billion. Finally, a relatively modest build-up of exports to 30 billion m^3 in new markets south and west of Turkmenistan could add almost US$300 million more in rent.

Table 2: Prospective Rents From Turkmen Natural Gas, Late-1990s

Market	Volume (b m³)	Cost (US$/mcf)	Unit Rent (US$/mcf)	Total Rent (US$ million)
Domestic	10	0.75	2.25	800
CIS	70	2.35	0.65	1610
Iran/Turkey	30	2.75	0.25	270
Total				2680

Source: Industry estimates (1997)

It is difficult to demonstrate the potential importance of such a rent stream for the Turkmen economy, given the complexities involved in making a meaningful estimate of its GDP during the transition to a market economy. But a rough indication can be gained if a number of simplifying assumptions are made and the potential viability of the ten year development plan can be assessed. If it is assumed that economic output had held up during the initial years of the plan and that the country's per capita income reached US$1500, the total GDP would be around US$7 billion, of which non-mining GDP would be around US$4.3 billion. The application of the user cost method of accounting for natural capital depletion when applied to those GDP estimates (under cautious assumptions about the gas reserves and interest rates), would allocate 36% of the natural gas rent to saving/investment and the remainder as true income (El Serafy 1989). This would boost the ratio of saving/investment by 22% of non-mining GDP, compared with an estimated investment level of 17% of GDP in 1996 (the latter is based upon 10% of GDP in public investment and around 7% private investment). The rent stream would also boost domestic consumption by the equivalent of almost two-fifths of GDP. Such figures are highly speculative, but they do suggest that a stream of rents on the scale envisaged by the ten year development plan might indeed have supported the gradualist reform policy that was intended.

The Slow Pace of Reform Under the Development Plan

Certainly, the first five years of the ten year development plan have been associated with tardy reform. Turkmenistan is one of the slowest reformers among the Central Asian components of the former Soviet Union, themselves a lagging group of countries. Privatization was pursued slowly in order to avoid both off-loading state assets at minimal prices and also the corruption associated with rapid privatisation in, for example, Russia. The development plan envisaged that most of the agricultural and mining sectors (comprising altogether, some 80% of economic activity) would remain in state hands until at least 1995. By 1995, the government claimed that, if 'home' industry and petty trading were taken into account, there were 21,000 private enterprises and 18% of GDP was in the private sector. The EBRD (1996), however, estimated that 91% of GDP remained in state hands: even commercial activity in the capital city had not been privatised. A bankruptcy law, introduced in 1992, has been little used and a competition policy had yet to be developed.

Between 1993 and 1996, only 1800 small state-owned enterprise (SOE) units (defined as employing less than 20 persons on average) were privatised, 40% by auction to private owners and the rest in the form of cooperatives. These small former SOEs comprised mostly services such as laundries and hairdressing, whilst food shops remained firmly under state control. As to the large SOEs (100 to 500 employees) which have caused such problems for another gradual reformer, China (*Economist* 1997a), only four out of the 600 which were targeted in 1996 for privatisation were actually launched. Elsewhere, six departments within the Ministry of Agriculture were to be sold off, but the terms reserved at least 51% state ownership. In rural areas, only small amounts of farmland were leased to private individuals prior to 1997, and although land ownership was legalised, a 15 hectare ceiling was set upon plot size (EBRD 1996).

Turning to public finance, central government expenditure had averaged 41% of GDP 1990–92, but that level halved in 1993 and halved again thereafter (although these figures exclude the activity of some ministries), before recovering back to 16% of GDP in 1996 (EBRD 1997). Turkmenistan ran budget surpluses during the early-1990s, thanks to a boost in receipts from natural gas exports. Two-fifths of the revenues came from taxes (including those from natural gas exports which accounted for 85% of non-farm foreign exchange earnings). The Turkmen government prudently accumulated foreign reserves which had reached US$1.17 billion by 1996, around one-half of nominal GDP. Taxes were reformed in 1992–93, cutting profits tax from 45% to 25%; introducing VAT at 20% (with some concessionary rates at half that level); setting income tax at a flat rate of 8%; and abolishing export taxes (EBRD 1996).

Monetary policy was effectively determined by the cabinet, given the country's rudimentary financial system, although control of credit officially functioned through bank reserve requirements and refinancing arrangements. The state bank became the central bank in 1993 with the launch of the national currency, the manat, and the remaining five largest state banks became creditors to fifteen commercial banks, some of which are owned by SOEs. But lax monetary policy caused the rate of inflation to accelerate sharply and average 1600% over the three years 1992–94. A stabilisation programme was then introduced. Inflation was expected to average 250% in 1996 on a declining trend, but this did not occur (EBRD 1996). Rather, the inflation rate remained just below 1000%, and a further acceleration seemed likely as a result of the reintroduction of directed credit in late-1996 (EBRD 1997). Interest rates were negative, and remained so, even after sharp rises in 1995. By 1996, Turkmenistan was the only reforming country which had still not brought the rate of inflation back to double digits (EBRD 1997).

In trade, the exchange rate was pegged to the dollar in 1993 but the government was forced to make periodic adjustments. The government decided to distribute foreign exchange through weekly auctions and placed trade restrictions on some imports. The tight restrictions on hard currency transfers repressed imports and impeded inward investment, even though the government was seeking to attract foreign investment in order to expand the production of natural resources. By 1995, import repression was adversely affecting production. Meanwhile, an export tax was replaced by a foreign exchange tax for the state sector which required that, from 1995, 50% of hard currency from state exports should be transferred to the central government (rising to 70% in the case of oil and gas). Increasing recourse was made to barter trade as economic conditions in the country's principal trading partners deteriorated, but this was banned in 1996 with the notable exception of natural gas (EBRD 1996). Severe import repression persisted, however, with increasingly adverse consequences for domestic production through 1996.

Labour reform proceeded slowly and labour markets remained far from flexible. Within the dominant state sector, wage rates were set by the government and they were raised in a uniform way across all enterprises, despite wide variations in skill shortages. A minimum wage, introduced in 1992, was quickly eroded by inflation. For example, when the rate was raised in 1996 it stood at only US$10 per month (EBRD 1996). Although official unemployment was estimated at barely 1.2% in 1993, production and employment (as well as exports from the energy sector) were subsidised. The transfers to SOEs took the form of bank credits, but in 1996 a change in policy occurred which sharply reduced this practice (EBRD 1996). Social security was funded, in part, by a 20% levy on wages and salaries. The labour force had few rights: work conditions in factories and on farms were often dangerous and unhealthy, while environmental damage has been substantial, whether in irrigated rural regions or industrial areas. Carbon emissions were 1.55 tonnes per capita in 1992 and total national emissions were 5.9 million tonnes (WRI 1996).

The net effect of the gradual reforms begun in the early-1990s has been to free up some prices, privatise a small part of the economy and attract a modest amount of foreign investment. But stabilisation has not been achieved, the financial sector is weak and state direction remains pervasive. The slow pace of reform was designed to cushion the populace from the upheavals of the transition to a market economy, but it did not prevent the per capita income from falling in 1996 to 57% of what it had been in 1989.

The Need to Rethink the Development Strategy

The Unravelling of the Hydrocarbon Export Plan

An initial upward adjustment of export prices in the early-1990s outstripped the adjustment of domestic prices, so that the ratio of exports to GDP jumped sharply from less than 10% to more than 50% (EBRD 1996, 207). This produced a strong positive trade balance which yielded sizeable foreign exchange reserves and formed the basis upon which the ten year development plan was drafted. However, contrary to the expectations of the planners, the trade balance narrowed after 1993 and was barely positive by 1996 according to estimates (EBRD 1997). This deterioration reflected a decline in exports as the nominal receipts on natural gas fell from US$1.8 billion in 1993 to US$1.02 billion in 1996. In fact, the actual revenues from natural gas were much lower than that by then. The government responded by repressing imports in order to conserve foreign exchange. GDP declined by almost half while the pursuit of a loose monetary policy triggered inflation.

The development plan was undermined by a combination of the loss of established gas export markets and the slow speed with which the external investment emerged to fund new pipelines. Far from expanding, the installed gas delivery capacity of around 80 billion m^3 per annum was not fully used. The actual gas deliveries dropped to one-quarter of existing capacity by 1994 as Gazprom denied Turkmenistan gas access to the European pipeline and payment arrears accumulated on sales to the Ukraine, Georgia, Azerbaijan and Armenia.[2] By 1996, some 53% of gas payments to Turkmenistan were in the form of barter goods, much of which proved of shoddy quality and therefore unsaleable. Meanwhile, of the 47% in cash payments, less than half was actually received and payments arrears reached US$1.4 billion.

As for pipeline construction, the early foreign investors in Turkmenistan's hydrocarbon sector were discouraged. The two most important early foreign investors were Bridas (Argentina) and Larmag (the Netherlands). Bridas alone invested US$400 million 1991–96 and planned to increase that sum seven-fold. However, both early entrants experienced problems with transportation which prevented them from expanding exports. Their problems intensified after the appointment of Kh. Ishanov as Minister of Oil and Gas in 1994 due to periodic cancellation of their export licences. Clauses were then introduced into their contracts which allowed the government to divert oil from export markets at prices set by government officials (RPI 1996). Such arbitrary discrimination by the new minister was discouraging to investors. However, it was targeted at the smaller companies with limited financial muscle: more appreciation was shown to larger oil companies such as Amoco and Unocal.

The slow expansion of investment, coupled with the deterioration in gas export revenues, prompted the Turkmen government to suspend gas deliveries in March 1997. But the importers responded by drawing down their gas reserves and turning to Russia for new deliveries. Fortunately for Turkmenistan, the IMF requirement that its CIS clients must not accumulate further debt by adding to their debt arrears, ensured an annual stream of $200 million to Turkmenistan from its indebted customer countries. But, even so, the loss of gas revenue after March 1997 was equivalent to a negative shock of around 10% of GDP for that year alone. And this shock occurred on top of the decline in real payments for gas and per capita incomes described earlier.

Meanwhile, government measures to boost food self-sufficiency and to construct factories in order to enhance the value added from cotton, back-fired. In 1992 the country had imported all of its sugar, almost three-quarters of its potatoes, 45% of its milk and one-third of its grain. The growing shortages of foreign exchange together with shortages of transportation and the disruption of shipping routes in 1993, encouraged the self-sufficiency strategy. While the reasons for pursuing food self-sufficiency are understandable on strategic grounds, such a strategy deflects farming capital and labour from their optimum allocation. Moreover, the transformation of most state farms and collectives into peasant associations in 1995 still provided insufficient incentives to elicit higher efficiency.

The results of these autarkic policies were far from successful. For example, the wheat area was expanded three-fold, but this entailed the use of marginal land while fertiliser inputs slumped. Meanwhile, the labour which was required per tonne of production increased and, since real wages simultaneously fell, farmers had a reduced incentive to exert themselves. Grain yields halved from 1.9 tonnes per hectare to only 0.9 tonne. Cotton production also fell sharply, to 430,000 tonnes in 1996 (one-third the 1992–94 level), because of pests and shortages of fertiliser, water and labour. Moreover, despite a rise in the procurement price for cotton, it was still barely 15% of the world price of US$1300/tonne (IMF 1997). Elsewhere, state instructions to increase the domestic cattle herd led to shortages of beef as farmers reduced the annual cull. The net effect of these unfavourable trends in farming seems likely to intensify the negative shock arising out of the cessation of natural gas exports to around 20% of GDP in 1997 alone. Such counter-productive policies make a clear case for price reform and reduced state intervention, in other words, a clear case for faster reform.

The deterioration in both production and exports turned the initial public sector surplus into a deficit which averaged 1.1% of GDP 1993–96. The government responded by accelerating the pace of reform towards the close of 1996. It introduced a land reform in December, but once again the change was cautious: private land rights were extended to collective farmers only, although other private citizens could lease land or be granted ownership at the government's discretion. However, land could not be sold and this sharply restricted the use of land for collateral. Nevertheless, those farmers who could demonstrate their capacity to manage farms satisfactorily over two years would be allowed to own their land and bequeath it to their offspring. In addition, farmers would be free to dispose of their entire production at market prices following plans to phase out state orders by the year 2000. At the same time, a further 2,000 SOEs in public catering and trade were to be auctioned off, but the state continued to specify profit margins in privatised firms (EBRD 1997).

Price controls were also reigned back so that those commodities with controlled prices and/or rationing declined from over 400 to around 50. But the latter still included basic food items, heating and housing and rations of gas, electricity and water continued to be provided free. The official price for bread was barely one-quarter the black market rate while meat was traded in free markets at thirty times prices in the rationed sector (EBRD 1997). But the ratio of prices on commodities that were controlled was subsequently allowed to rise sharply towards the average price level.

Restructuring the Development Plan

The deterioration in the markets for Turkmen gas mean that the government has failed to realise the potential of its resource endowment under the ten year development plan. The country's per capita income had been only US$980 in 1993 but it fell to US$530 by 1996 (EBRD 1997). However, it should be noted that a combination of price repression and cheap labour means that this low per capita income translates into US$2,500 in purchasing power parity terms (*Financial Times* 1997). Nevertheless, a change of development strategy is clearly required and agriculture appears to offer the best immediate prospect for resuming economic growth. The adverse impact on the economy of a very poor harvest in 1996 underscores the need to secure the potential benefits of a more dynamic performance from this sector. In contrast to farming, the contribution of the mineral sector should be expected to strengthen through the long-term because mineral-driven development is capital-intensive and its investment has long lead-times. The delayed impact is likely to be compounded in the case of gas-based development by the fact that pipeline construction is difficult and uncertain in a politically fragile region.

There is strong evidence that an efficient agricultural sector plays a crucial role in the early stages of economic development (Mellor 1995, Tomich *et al.*, 1995). The sector initially acts as an important source of foreign exchange and taxation and it also provides inputs for 'early industry' such as food processing and textile production. The agricultural sector also provides an important cushion against unemployment by absorbing surplus labour until such times as the emergence of labour shortages in the economy stimulates increased farm productivity (Auty, 1994). However, in many resource-rich countries, the potential of agriculture has been damaged either by neglect of the needs of potentially efficient small farmers, or by the introduction of crop marketing boards and price controls which squeeze margins and eliminate production incentives.

A pre-requisite for the revival of the agricultural sector is to permit specialization in the production of crops which provide the highest returns. This calls for the break-up of collective and state farms into smaller units and for the abandonment of food self-sufficiency. This is because there are few economies of scale in agriculture, with the exception of some agri-business operations concerned with fresh produce and also those crops where immediate processing is required in scale-sensitive plants. Even where centralised processing equipment and research facilities are required, it is invariably more efficient to split the crop production unit (the farm) from the processing unit (the factory), a system known as the 'nucleus plantation' which Malaysia has adopted with success (Graham and Floering 1984). Elsewhere, Binswanger and Deininger (1993) and also Deininger and Binswanger (1995) argue persuasively that countries which have historically suppressed yeoman farming systems in favour of larger 'yunker' estates or collective holdings, have been obliged to provide significant subsidies to compensate for the resulting inefficiencies.

If agricultural production is to be improved, more efficient use must also be made of water. Historically, the water resources of Turkmenistan have been badly managed. By 1989, Turkmenistan consumed 6,400 m^3 of water per capita in 1989, 50% more than the next highest user in the CIS, four times North American levels, and ten times EU levels (EBRD 1996). The wasteful use of water resources was caused in part by the country's excessive pre-occupation with cotton production. The rapid construction of the 1000 kilometre Kara Kum canal which brings water from the Syr Darya River neglected to line the canal bed. As a result, the rate of water losses due to seepage and evaporation combined is estimated at 70%. Meanwhile, the careless application by farmers of water, which is provided free of charge, has led to problems of salinization. A rational pricing system is therefore required in order to: allocate water supplies to sectors giving the highest return; encourage thrifty use; and generate revenues to improve the water distribution system. Applications of fertiliser and pesticides have been similarly excessive due to the absence of sound pricing policies.

Government efforts to force the pace of industrialization during the ten year development plan have been counter-productive. Some 74 medium- and large-scale textile enterprises were established, employing 18,000 workers in the production of fabrics and clothes. Although the factories were expected to boost the value added per kilo of cotton by three times in the case of fabrics and six times in the case of clothes, reports suggest that the plants are not efficient. In fact, it is likely that the expanding private sector offers better long-term prospects for industrial growth than the state sector does. But even so, manufacturing in Turkmenistan is initially constrained by a lack not only of the required skills but also of good access to a large market as well as imported inputs. The latter problem may be best resolved by establishing an Export Processing Zone (EPZ). Yet, the landlocked location of Turkmenistan does afford some 'natural' protection for domestic manufacturers, as well as scope within the group of 'early industries' such as food processing and textiles, for local production. Construction materials, cement and petrochemicals should also prosper through the medium-term.

An attractive medium-term prospect for the Turkmenistan hydrocarbon sector is to boost exports of oil. One reason for this is that, as noted earlier, oil has much lower transport costs per unit of heat value than natural gas and, therefore, it yields higher rents. For example, even oil barged across the Caspian Sea (and therefore securing few benefits from the economies of scale) could earn a rent of about US$9 per barrel (some two-fifths of the revenue). In line with this more attractive hydrocarbon option, Mobil has begun investing with Monument of the UK in oil fields in the western part of the country. Mobil plans to boost oil output by 200,000 bpd in order to justify the construction of new oil pipelines via Iran, Kazakstan or the Caspian Sea (*Financial Times* 1997). Elsewhere, Amoco has replaced Bridas in some oil fields while the Argentinean company pursues legal battles over its initial contracts in the international courts. Meanwhile, the Turkmenbashi refinery is being upgraded with a $580 million investment.

The government of Turkmenistan remains committed to a process of gradual reform and China may offer an appropriate model. This is because of the dominance of agriculture in both countries on the eve of reform and the speed with which output in that sector can respond to policy shifts. Moreover, much of Turkmen industry is likely to be unattractive to either foreign or domestic investors, so that progress in both the mining and the manufacturing sectors is likely to depend on new greenfield investment which will take some time to build up. This would suggest there is a strong case for setting up an EPZ in which the geographical concentration of sound infrastructure and the minimisation of red tape can speed up the provision of attractive conditions for new domestic investment. Such a concentration of new investment would have the added advantage of facilitating the monitoring and control of pollution emissions.

The most critical determinant of the pace of reform is, however, likely to be the degree to which the accumulated foreign exchange reserves, estimated at US$1.17 billion in 1996, will allow the government to continue funding those less dynamic sectors of the economy which it has sought to insulate from change. If the demands of the sheltered sectors prove too large and the country is forced to seek external financial assistance, then slow reform may cease to be an option. Foreign debt had already reached US$680 million in 1996, some 30% of GDP, and debt service was estimated to absorb 18% of the country's export earnings. It is therefore difficult to see how present conditions can be sustained in the absence of a rapid reversal of the decline in export earnings, principally from natural gas. Current IMF proposals for a significant expansion of special drawing rights may prove timely for the besieged Turkmen economy.

Conclusions

The government of Turkmenistan opted to use the country's vast natural gas deposits to permit a gradualistic approach to social and economic reform. But the country lacks access to markets and its landlocked location combines with the potential hostility of neighbouring countries to render the construction of transport links risky as well as expensive. Although plans to reach markets in East Asia and South Asia seem unlikely to come to fruition in the near future, a route via Iran to Turkey and beyond to the EU, offers medium-term prospects for expansion. Even then, the high costs of transporting natural gas will absorb the bulk of the revenues, leaving a relatively low fraction of the revenues as rent. The rent is likely to be less than one-sixth of the gas price compared with around three-fifths of the price on more easily transportable oil. Worse, the collapse of traditional gas markets has inflicted a sharp negative shock on Turkmen GDP which fell by 43% 1989–96 and seems likely to decline by a further one-fifth in 1997.

The pursuit of a strategy which seeks to use the country's large natural gas deposits in order to transform Turkmenistan into a second Kuwait is no longer practical, however. But the slow pace of economic reform stifles the contribution which the agricultural sector can make to improved productivity and rising incomes. The development of the agricultural sector calls for an acceleration of privatisation, something which the government began in December 1996. The new strategy will also require the maintenance of incentives to farmers while realigning the prices of both farm products and farm inputs (notably water) in line with their relative scarcity. In the absence for some years of either new emigration opportunities or the achievement of a sustained rapid rate of labour-intensive economic growth, the expanding workforce will exert downward pressure on wage rates and thereby heighten income inequality in any reformed and flexible labour market. This is because the workforce is expanding at almost 3% per year, despite a deceleration in the rate of population growth to 2.3% per annum 1990–95 (WRI 1996).

By late-1997, the sizeable foreign exchange reserves which the Turkmen government had accumulated from the early-1990s assumed critical importance. The reserves can help adjust to the large negative shock which the economy experienced in 1997. But in the absence of a strong rebound from agriculture and gas exports, it is likely the country will need to request external assistance. At that point, slow reform may cease to be a practical option, whatever its merits.

References

Auty, R.M. (1990), *Resource-Based Industrialization: Sowing the oil in Eight Exporting Countries,* Clarendon Press, Oxford.

Auty, R.M. (1994), 'Industrial policy reform in six large newly industrializing countries: The resource curse', *World Development,* 22, pp. 11–26.

Binswanger, H.P. and Deininger, K. (1993), 'South African land policy: The legacy of history and current options', *World Development,* 21, pp. 1451–75.

BP (1996), *BP Statistical Review of World Energy 1996,* British Petroleum plc, London.

Deininger, K. and Binswanger, H.P. (1995), 'Rent-seeking and the development of large-scale agriculture in Kenya, South Africa and Zimbabwe', *Economic Development and Cultural Change,* 43, pp. 493–522.

EBRD (1996), *Transition Report 1996,* European Bank for Reconstruction and Development, London.

EBRD (1997), *Transition Report Update 1996,* European Bank for Reconstruction and Development, London.

Economist (1997a), 'The long march to capitalism', *The Economist,* September 13, pp. 23–26.

Economist (1997b), 'Turkey and Iran', *The Economist,* August 2, 46.

El Serafy, S. (1989), 'The proper calculation of income from depletable natural resources', In: Ahmad, Y.J. El-Serafy, S. and Lutz, E. (eds.) *Environmental Accounting for Sustainable Development,* World Bank, Washington DC, pp. 10–18.

Financial Times (1997), 'Turkmenistan belatedly joins the queue for oil and gas investment', February 14.

Graham, E. and Floering, I. (1984), *The Modern Planation in the Third World,* Croom Helm, Beckenham.

IMF (1997), *Turkmenistan: Recent Economic Developments,* IMF, Washington DC.

Mellor, J. (1995), *Agriculture on the Road to Industrialization,* Johns Hopkins University Press, Baltimore MA.

RPI (1996), 'Duke of oil', *Russian Petroleum Investor,* August 6, pp. 38–42 and 53.

Sagers, M.J. (1994), 'Long-Term program for Turkmenistan's Oil and Gas Sector', *Post-Soviet Geography*, 35 (1), pp. 50–62.

Stauffer, T. (1975), *'The prospects of energy-intensive industry in the Persian/ Arabian Gulf'*, Mimeo, Centre for Middle Eastern Studies, Harvard University, Cambridge MA.

Tomich, T.P., Kilby, P. and Johnston, B.F. (1995), *Transforming Agrarian Economies: Opportunities Seized, Opportunities Missed*, Cornell University Press, London.

US Department of State (1994), *Turkmenistan Economic Policy and trade Practices*, US State Department, Washington DC.

Wood, A. and Berge, K. (1994), 'Exporting manufactures: trade policy or human resources?' *IDS Working Paper 4*, Institute of Development Studies, University of Sussex, Brighton.

World Bank (1996), *World Development Report 1997*, World Bank, Washington DC.

World Bank (1997), *Commodity Markets and the developing Countries*, 4 (2), pp. 37–38.

WRI (1996), *World Resources 1996–97*, World Resources Institute/Oxford University Press, Oxford.

Note: This research was sponsored by the Harvard Institute for International Development under a cooperative agreement with the United States Agency for International Development.

Notes

1 Turkey was anxious to reduce its dependence on Russia for natural gas exports, and the Turkmenistan link opens up the possibility of Turkey securing gas from that country rather than from Iran (*Economist*, 1997b). The alternative sources for Turkey are Algerian LNG (US$3.50 per mcf) and Egypt (via an Israeli pipeline) for slightly less, but both these options carry significant political risk.

2 Gazprom is believed to wish to cut the border price demanded by the government of Turkmenistan from US$42/mm^3 down to US$28. This would allow southern Russian markets to receive gas at a delivered price of US$60. Gazprom would then be able to substitute its own gas for Turkmenistan gas in the Ukraine market where it could realise a price of US$80.

Chapter 12

The Demographic Boom and its Impact on the Mountain Regions of Tajikistan

Khojamakhmad Umarov

Introduction

The Central Asian republics had the highest population growth rates in the USSR. In 1987 they were estimated as follows. Uzbekistan: 31.1 per 1,000 per annum; Kyrgyzstan: 25.3 per 1,000; and Tajikistan: 34.9. Rates of this order had been maintained throughout the previous three decades, and so far they have shown no sign of levelling off.

The current demographic situation in Central Asia has resulted in a large complex of interrelated problems, which have exerted an increasing effect on the social and economic development of Central Asia, and even further afield. One of these problems is the increasing demographic pressure on mountain areas. We may take Tajikistan as a case study, but this process is also taking place in Kyrgyzstan and Uzbekistan.

Population Increase in Mountain Areas

In Tajikistan, the population is growing faster in the mountains than in the lowlands. Over the period 1970–1987 the Republic's population grew by more than 40%. In the valleys, the increase was 38%, while in the uplands it was 52%. The population increase in the valleys is due to natural growth, but in the majority of mountain areas, 20% to 30% of the growth is caused by immigration. These are the regions whose populations were moved in the 1930s to the lowlands and river-valleys of the Republic to establish large-scale cotton farms. Today, this process is being reversed: people are migrating from the cotton-growing valleys to the mountains where their parents and grandparents came from, because of the rapid

186

increase in population density and lack of space in the valleys. In the countryside around Dushanbe and Hissar the population density is as high as 1,400 people per km^2. Between 1970 and 1979 the area of irrigated land per person in Tajikistan shrank from 0.18 to 0.14 hectares. The total area of arable land per person (including almost all kinds of cultivation) fell from 0.26 to 0.19 ha. According to agriculturalists, the area of land necessary today to supply one person's food needs is 0.5 to 1 ha.

The modern migrants are hoping to find enough space in the mountains to be able to improve their living standards. Based on our observations, over 500 *kishlak* (villages) in Tajikistan's mountain gorges, deserted during the previous wave of migration, have been repopulated in recent years. Forecasts suggest that this process will accelerate over the next ten years, since the low level of development and poor standards of living in urban areas mean that the rural lowland population is unlikely to migrate to the towns and cities in large numbers.

Effects of Population Increases in Mountain Areas

On the whole, this process is a positive one, since it has created opportunities for the mountainous regions greatly to increase their food production, and contribute to the solution of food supply problems elsewhere. Suffice it to quote only one example: the design and planning institute Tajikgiprozem has estimated that the foothill regions contain over 734,000 ha of lands which are potentially suitable for vinegrowing. There are also considerable opportunities to develop upland soft-fruit and vegetable production, and to set up large specialised farms to grow nuts: walnut, almond, pistachio, and so on.

However, the population increase in the mountainous regions has also been causing some justified concern. The process of migration is disorganised, with the result that in most places, despite the *de facto* creation of settlements, elementary social infrastructure is non-existent: there are no schools, libraries, electricity, or other social services.

The immigrants also pose a threat to the delicate mountain environment. Forty years ago, a person living in the mountains was at one with nature; thousands of years of experience had taught him to aspire to preserve a balance between his economic needs and those of the natural environment. Recent years have seen a radical change in man's ecological role in the mountains, and his potential for destruction has multiplied. The forced resettlement of mountain-dwellers in the hot river-valleys, and the complete change in living conditions they experienced there, broke their links with this long-established mountain way of life. Their contacts with the mountains became 'consumerised'. They are now returning to their forefathers' homes armed with electronic equipment, bulldozers, graders and dump-trucks, excavators and compressors, explosives and chemicals, high-precision shot-guns and industrial fishing-nets, and motivated by insatiable demand. The incomers are completely ignorant of how to behave in the mountains: they have built wide roads along many mountain gorges and even on small river-banks, using powerful machinery and explosives; and in some areas felled and uprooted trees 100 to 300 years old for firewood. The numbers of wild goats and sheep have fallen to such an extent that these animals have been declared endangered species.

It would be unfair to attribute all this degradation to changes in the ecological behaviour of the rapidly increasing population in high altitudes; it must be considered from the point of view of the overall regional demographic and economic situation.

Effects of Economic Development on Mountain Areas

Demographic and economic factors are closely interrelated. Central Asia's economic growth over the last thirty years has been characterised by the increasing proportion of economic activity which takes place in industries which are capital- and raw material-intensive. This trend, combined with the population increase, has contributed to a narrowing of employment opportunities proportional to the able-bodied population, and an increase in surplus labour. In other words, the region's economic growth has been in the form of extensive development. This trend has had a direct effect on the mountain areas (through the establishment there of hydroelectric power engineering, mining enterprises, health-care services, sanitation etc.), as well as an indirect effect, through the methods used to solve development problems.

A vivid example of such a direct effect is the construction and exploitation of the Norak hydroelectric power station and reservoir. Here, 6,800 ha of fertile land have been submerged by flooding around the reservoir, whose banks are now rapidly being eroded altogether. The process of clearing away the water and lowering its temperature has been slowing down the growth rates of cotton and other crops.

The more general effects of this kind of economic development also seem not to have helped Central Asia. Living standards remained lower in the Central Asian republics than in most of the rest of the USSR. Moreover, Central Asia was falling further behind, as the following table illustrates.

Percentage increase in monthly wages, 1970–1987

	Whole Soviet Union	Tajikistan	Uzbekistan	Kyrgyzstan
Industrial, office and professional workers	66.3	41.0	47.8	52.2
State farm workers	98.9	58.0	67.4	82.4
Collective farm workers	130	60.2	43.1	93.1

It should also be borne in mind that families are larger on average in Central Asia, and contain a higher proportion of dependents. These facts have led to increases in the numbers of individual farmsteads in Central Asia, and increases in livestock numbers.

Percentage increase in livestock, 1970–1987

	Whole Soviet Union	Tajikistan	Uzbekistan	Kyrgyzstan
Cattle	21.6	30.0	41.8	22.2
Sheep and goats	2.7	26.9	6.2	10.6

As a result, the density of livestock in the mountain areas has long since overstepped the set limits, and now exceeds them by 2 to 2.5 times. During the spring and summer seasons all sheep and goats and a considerable proportion of cattle now graze on the summer mountain pastures. Overgrazing has resulted in the degradation of the hydro-physical properties of mountain soils and an increase in erosion. The productivity of Tajikistan's summer pastures has fallen by almost half in the last twenty years. In many gorges networks of tracks cover up to 56% of the mountainside.

The high population growth rate, low living standards and land deficit in the mountainous parts of Tajikistan have inevitably encouraged intensification of the rural economy, with increases in the use of synthetic fertilisers, chemicals and machinery. The results are regrettable. Pesticide residues have already been detected in soil and plants near the high-altitude Lake Zorkul in the eastern Pamirs. Extensive irrigation of mountain slopes has resulted in the speeding-up of soil wash-off, which in some areas has reached precarious dimensions. As a result, the proportion of mountain soil area which is eroded has now reached 72% in Tajikistan, 82% in Uzbekistan and 92% in Kyrgyzstan.

The Role of Management and Administration

The reasons for the deterioration of mountain lands are diverse, and have been the subject of scientific investigation. But very little attention is given to the principal cause, the most powerful destructive force for mountain areas. This is the indirect effect of economic development referred to above, the command-administrative management system, which governed all spheres of social life, and has now held sway over nature and natural resources for decades. The command management system has had a negative effect on mountain regions in the following ways:

1 Planning targets and indices of success have been rigid, unbalanced and scientifically ungrounded. Very often there was no coordinated relationship between the objectives set by the state and the means of achieving these objectives. Silk-worm breeding in the mountains may serve as an example. The targets for the production of cocoons were increased from year to year, while reserves of forage for the worms fell far below real requirements. The industry was not supplied with the necessary resources, particularly in energy. As a result, intensive degradation of mulberry trees has been observed; a drastic cut has taken place in the production of mulberries, once one of the most important foodstuffs of the inhabitants of the region; and the felling of trees for firewood has reached excessive levels in the areas where silk-worms have been bred.

2 Local organisations were ignored and humiliated. Organisations such as village and district Soviets of People's Deputies, Forestry Committees, etc., were responsible for the state of affairs in the mountain territories. State projects to build hydroelectric power stations, reservoirs and open-cast mines, and to increase the number of livestock on high altitude pastures, did not take into consideration the opinions of these organisations. These local bodies expressed the interests of the local population, and had a

mission to monitor the ecological balance in the mountains. Large projects in mountain areas fulfilled the interests of government departments, which were often claimed as public interests, but were not necessarily those of the local population or environment.

3 The command management system gave free rein to the dogmatism of bureaucratic managers, in thinking and economic and ecological behaviour. This led to the natural resources of mountain territories being used in an irrational, wasteful way. Official dogmatism was expressed particularly in the unthinking transfer to mountain areas of those forms of farming which had been adopted in the lowlands. For example, the process of setting up collective and state farms in the mountain territories completely ignored the protection of natural resources. Any change introduced from the top down in the agricultural development of lowland areas was automatically extended to farms in the mountains. The campaign to amalgamate collective farms waged in the 1950s involved mountain farms as well as lowland ones. The same is true of the campaign to concentrate and specialise agricultural production on the basis of inter-farm cooperation and agro-industrial integration. This policy led to the emergence of large livestock-breeding complexes in the mountain regions, which created a threat to vegetation and wildlife, and drastic water pollution. Experts have revealed a serious deterioration of natural vegetation cover on the southern slopes of the Vakhsh range and the north-western slopes of the Hazratshoh range, between which lies the largest cattle-breeding agro-industrial complex in Central Asia. This has brought about a drastic fall in the forage productivity of pastures, which now amounts to a loss of 40–50% of total herbage. The rapid increase in cattle numbers in this enterprise resulted in underground water pollution all over the territory of the complex, and a sharp increase in hydrogen content in all the small springs along the river Obi Mazor. Such changes show how dangerous stereotyped attitudes to economic development can be for mountain territories.

Today, the economic development process in Tajikistan is changing from administrative-command methods to economic means of management. Nevertheless, this transition will not automatically harmonise the relationship between economics and the environment. This task is not an easy one, and involves much more than the execution of a wide range of measures to rehabilitate ravaged mountain areas. The economic methods of management applied to these territories must maintain a rational and proper correlation between production and the environment. These relations have to acquire an organic nature.

A Mountain Development Strategy

All the factors mentioned above argue in favour of the elaboration of a scientifically well-grounded development strategy for mountain territories. This strategy should include both general and particular elements. It should take into account the serious problems which are common to mountain regions all over the world: in this respect such a strategy would have a global aspect. But a mountain development strategy will only bring about results if it is concretely applicable in particular regions, taking into account all the factors which could stabilise relations between society and nature in those regions. Key tasks would be the rehabilitation of ravaged areas of mountain landscape, forest cultivation, restoration of grassland, etc. The ultimate goal of such a strategy would be 'ecological-social-economic' development of mountain areas.

Measures to realise a development strategy for mountain regions should be based on the joint efforts of industry and agriculture, the population and various institutions of both plains and mountain regions. This collaborative principle is particularly justified in the conditions of former-Soviet Central Asia, since those anthropogenic changes which have taken place in its mountain areas were caused by the requirements of economic and socio-demographic development (irrespective of whether they were rational or irrational). Hence, the whole of society, having benefited from this development, should participate in repairing the damage. Enterprises and institutions, republican and international bodies should coordinate their efforts, and share the costs. The most necessary condition for the success of such collaboration is a clear definition of the roles of the various enterprises, departments and other bodies. For example, the efforts of collective farms and lowlanders who graze cattle on high-altitude pastures should be directed not only to rehabilitating eroded pastures, but also to carrying out other measures to increase the forage available. International budget resources should be used for measures of CIS-wide and global importance, such as the prevention of intensive melting of glaciers.

The development strategy for mountain territories will need to be complex. The technological, social, political, economic and organisational aspects of the strategy should be distinguished. In Central Asia demographic factors will play an important role, but since they are relevant to all these aspects, there is no need to separate them into an independent subsection of the strategy.

The technological aspect of the strategy should proceed from the methodological principle that an ecological and economic balance in mountain areas will be possible only when new productive forces are created which give primary consideration to the specific features of high-altitude regions. The question concerns the particular productive forces of mountains. So far, these consist of the plants and wildlife which have evolved there over millions of years. Everything that has penetrated into these territories during the industrial revolution constitutes strong competition for the mountains' natural productive forces, and has therefore favoured their rapid degradation. Thus the vital task is gradually to clear mountain territories of those newly introduced productive forces which are alien and destructive. Less important is the problem of how to extend the natural productive forces (e.g. rehabilitation of pastures, meadows and hay-fields), since the domestic and wild fauna have adapted optimally over the centuries to the specific conditions of high altitudes.

But the task of utmost importance is the development of machinery and appropriate technology which is ecologically safe and maximally adapted to particular mountain regions. The same applies to methods of farming, transport, road construction, etc. Machinery and infrastructure must be designed for the conditions of Central Asia, taking into consideration the demographic load on the mountain areas.

The mountain development strategy should promote positive solutions for the most important social problems which originate in both mountain and plains regions, notably, overpopulation. Academician N. N. Moiseyev is quite right when he argues that 'any surplus population in environmentally-sensitive regions is in itself a risk'. First of all, an ecological and demographic evaluation of mountain areas should be carried out to determine the carrying capacity of each area. Then, appropriate conditions should be created for the control of population growth. Family planning methods which go against the moral values of the population are inhumane. It is therefore necessary to raise the cultural level of the people, provide education and full employment, and promote urbanisation in mountain regions. Finally, measures must be taken to restrict migration into areas where the demographic load is critical.

The political aspect of the strategy is principally concerned with the relationship between autonomy and dependence in mountain regions. As a part of the state they are to a certain extent dependent and their functions are defined by this position. At the same time, they must also have some autonomy, for without it, realising their vital interests and those of their population will be seriously complicated. B. Messerly is right when he notes that 'as yet optimal relationships between dependence and autonomy in mountain regions have not been found'.

The economic aspects of a mountain development strategy should first of all define relations of production which foster the maintenance of an ecological and economic balance, effectively imposing sanctions on those who endanger the environment. In the conditions of former-Soviet Central Asia the establishment of such relations should take into account the requirements of neutralising demographic pressure. This can be achieved by limiting human activities in ecologically vulnerable mountain areas, by such means as the extension of nature reserves and creation of national parks. In the author's opinion it would be generally expedient to turn forestry enterprises into cooperatives. These would operate on the basis of state orders. It would also be reasonable to establish a whole network of nature protection cooperatives, dealing with conservation of glaciers, natural monuments, protection of reserves, etc. Long-term leasing of pastures could also be justified.

Radical changes are needed in the economic relations between the mountains and the plains, between their enterprises and organisations. Thus, for example, the Norak hydroelectric power station, situated in the mountains, is the most profitable enterprise of Tajikistan's national power system Tajikglavenergo. Nevertheless, in spite of the fact that its functioning is associated with flooding and other negative consequences, it does not allocate any resources for environmental protection.

We can only hope that political changes following the collapse of the former Soviet Union will lead to the establishment of better relationships between the mountain regions and the rest of society.

Chapter 13

Promoting Integrated Mountain Development in the Hindu Kush-Himalayas

Mahesh Banskota

The strong interrelationships between environment and development in mountain areas need no further reiteration. The major problem is still achieving the systematic integration of environment and development in day-to-day decision-making processes. The practical problems involved are complex and the resources and institutional capacity needed for this integration are very large indeed. The prevailing conditions in mountain areas, with regard to both the economy and the environment, leave little room for optimism. Rapid population growth, deforestation, economic stagnation and institutional failure are widespread. The difficulties associated with meeting both the immediate needs of the population for food, energy, education, health and shelter, and the requirements of long-term environmental sustainability are formidable. This is further complicated by the fact that political preferences are always weighted towards overcoming short-term problems. The need for a change is obvious: simply doing more of the same in decision-making, choice of technology, and management systems is unlikely to move mountain environments and economies towards a sustainable path.

There are some positive conditions. In spite of the growing influence of development activities in mountain areas, the majority of mountain people are still highly dependent on renewable natural resources, rather than, as elsewhere, on non-renewable ones. As industrial growth is in a very early stage, environmentally preferable options can be introduced more readily in mountain regions than in other areas, where change is more difficult, due to the large investments already made. In many mountain areas, the abundant sources of clean, renewable hydroelectric energy provide immense potential for development. With few other readily available exports, tourism has great potential for development, and could provide a critical source of foreign exchange. The history of local institutional innovation, particularly in natural resource management, is also very rich in mountain areas and stronger efforts are needed to mobilise this experience. While these are a few examples of potentially useful options in a situation which appears dominated by the many adverse conditions, the challenge is to develop these potentials systematically and on a long-term basis. This challenge demands a very strong process of integration, firmly based on a proper understanding of the critical links between environment and development.

The International Centre for Integrated Mountain Development (ICIMOD) was established ten years ago with the objective of promoting integrated development in the mountain areas of the Hindu Kush-Himalayas (HKH). With so many bodies (governments, non-governmental organisations and international non-governmental organisations) already active in development, the need was felt for an organisation that could analyse and synthesise development issues and experience, and identify better approaches for integrated mountain development. This chapter briefly explains the activities of ICIMOD and the types of mountain development problems it deals with. Section 2 provides an overview of mountain development in the context of the HKH region. Section 3 outlines the history of ICIMOD. Section 4 assesses ICIMOD's programme experience, and the last section identifies some lessons for the future.

The Context of Mountain Development in the Hindu Kush-Himalayas

Many parts of these mountains have passed through different stages of interaction between population, agriculture and the environment. Despite many complex local variations, these stages reflect the typical evolution of an agricultural economy. In the earliest stages (quasi-subsistence) population densities were relatively low in relation to available agricultural land. Agricultural practices were extensive, with widespread shifting cultivation across marginal lands where irrigation was difficult. Labour utilisation rates were low and technology was more or less static. Demand for non-agricultural goods was limited to some basic items, much of which were home-produced. Urban centres were few and far apart, performing largely non-economic, surplus-extracting functions. Incentive to trade was low because of low agricultural productivity and the limited availability of non-agricultural goods, except in areas which were unable to grow their own food and had to trade with other regions. Over-extraction of natural resources from the environment was not extensive and the overall balance of demand and availability of natural resources was relatively favourable. These low-productivity, near-equilibrium conditions did not last very long. Many quasi-subsistence pocket economies quickly deteriorated. Human populations increased, at rates close to 2% a year. Livestock numbers increased similarly, to provide essential food. In many cases terracing, the development of irrigation, integrating livestock into farming systems, and even switching over to crops from the New World, such as maize and potatoes, gave significant relief for varying periods. Out-migration also provided a temporary safety net. But it was only a matter of time before an increasingly large number of mountain pocket economies became locked into a Malthusian trap. The prevailing inequalities in access to natural resources, the effectiveness of local resource regulations and access to wider markets all played important roles in determining the outcome of the growing imbalance between demand and availability of critical natural resources.

Important changes have come from within the community in response to these pressures on the economy and the environment. Mountain farmers' adaptive strategies have been the most important agents of change in the past and will continue to play a crucial role in the future sustainability of mountain agriculture. Farmers have had to adapt to many forces, such as limited land and forest resources, population pressures, increased market penetration, crop diseases, soil erosion, etc. Abandoning terraces, increased cropping intensity, expansion of horticulture, planting of fodder trees, stall-feeding of livestock and out-migration are some of the steps they have taken.

Changes from outside have also played an important role, with improvements in access and increasing public sector programmes in mountain areas. The mountain farmer is now beginning to receive a whole range of support from outside. While large parts of mountain areas still lack access to outside support, in areas where this is being provided, hill farmers have been willing to take major risks to diversify and change traditional farming practices and undertake new off-farm activities such as catering to tourism, carpet-weaving and seasonal migration to the cities.

Improvement in accessibility has brought the hill farmers closer to larger markets and encouraged them to switch production gradually from subsistence crops to high-value commercial crops. Once markets are assured, the changes are far-reaching. Research and extension services have helped to bring the green revolution to mountain areas by introducing suitable kinds of high-yield variety seeds to the lower-lying districts.

Another major source of transformation of mountain agriculture has been through institutional development. Mountain societies have traditionally sustained relatively autonomous institutional mechanisms. These mechanisms were, furthermore, closely focused on the regulation and management of common property resources such as forest, water and pastures. In some groups they also regulated the use of agricultural lands. With the increasing role of the state (in terms of military activities, revenue collection and more recently, development activities) and the market, the influence of traditional institutional mechanisms started to weaken. Outside decisions, external products and technologies began exercising greater influence on local resource use and the availability of economic opportunities. Local organisations were progressively replaced by state agencies, completely altering the traditional system of regulation and use of common property resources. These processes led, in many cases, to the slow demise of local organisations, and in other cases to their substantial weakening.

Some of this is slowly changing as disenchantment with national agencies grows, and efforts are being made to promote local organisations. Usually, these efforts come from external agencies, and their major thrust is to mobilise and support farmers' own initiatives to resolve their problems. While this should still be seen as a learning process, the successes achieved through such institutional innovations have been very encouraging.

Despite mountain farmers' very long history of developing many different types of adaptive strategies and institutional innovations, the overall scenario for mountain agricultural economy appears very bleak indeed, given the overall trends in almost all the key resources. Some of these trends may be summarised as follows:

i) *Cultivated land:* Declining fertility; loss of land; increasing soil erosion and partial desertification; major problems of water management.

ii) *Pastures:* Large-scale overgrazing and degradation of pastures; landslides; livestock management focusing more on numbers than on quality.

iii) *Forests:* Reduced forest area; decreasing forest crown cover; continuing encroachment by overgrazing; loss of species; limited government control on wood harvested; very poor afforestation.

iv) *Water:* Increasing shortages; flash floods.

v) *Households:* Large households; more dependents; labour scarcity; increasing burden on women; higher proportion of children in the labour force; reduced cultivated area per capita; increasing landlessness; frequent and permanent food deficits; scarcity of firewood; indebtedness; outmigration.

vi) *Economy:* Rising prices for food, energy and other inputs; political pressures for greater subsidy on inputs; decline in non-agricultural activities due to further concentration on food production; investment meets short-term needs at the cost of long-term, sustainable development; difficulties in mobilising domestic resources; increased dependence on external inputs and resources.

vii) *Environment:* Increase in soil erosion; increasing danger of destruction to roads, bridges, terraces, hydroelectric plants; greater incidence of floods; changes in micro-climate; negative downstream effects.

viii) *Social characteristics:* Very low literacy levels, even lower for women and in rural areas; life expectancy improving but still low; infant mortality high. Burden on women to meet the subsistence needs of the household increases as forests diminish, agricultural productivity declines, water sources dry up (making distances to water sources greater) and men outmigrate.

The overall socio-economic situation in the mountain agricultural economy and environment is fairly distressing. Poverty is becoming more firmly entrenched as available natural resources degrade, population demand escalates and development activities produce little impact.

The Establishment of ICIMOD

It was in response to the deteriorating mountain environment, and the difficulties of moving development forces uphill in a rugged terrain, with a scattered population and wide micro-environmental diversity, that the need for a better understanding of the mountain environment and economy, and of relevant development experience was considered urgently needed. The prevailing development strategy, based on industrialisation and designed for plains and cities, needed urgent reexamination, and an approach more appropriate to mountain areas needed to be developed. It was widely recognised that there was a huge pool of knowledge on different aspects of mountain development in the experience of development projects, and that this needed to be examined and synthesised. National approaches to the development of their respective mountain areas needed to be evaluated, so that the conditions which led to successes and failures could be identified. More importantly, the historical isolation of policy-makers and development workers had to be broken, through information exchange, collaborative research and training and, most importantly, by establishing a mountain development forum. It was against this background that ICIMOD came into existence.

The problems outlined above were discussed in the 1970s and it was concluded that, considering the complexity of the issues, a holistic and integrated approach to development was needed. There were also strong recommendations for the establishment of a regional institution which would focus on integrated mountain development, covering both environmental and development aspects, and cutting across all sectors. In addition, the focus of this institution was to be on the Hindu Kush-Himalayas, the longest, highest and most populous mountain chain in the world, which was experiencing severe economic and environmental problems. Extending 3,500 km from east to west, the region had a population of over 120 million and contained a wide variety of ecosystems, from the subtropical uplands of Myanmar (Burma) and the Chittagong Hill Tracts to the cold, dry mountains of Xizang (Tibet) and the warm, dry mountains of Afghanistan.

The proposal for a regional institution was discussed and endorsed at the regional UNESCO Meeting in 1975, when the Kingdom of Nepal offered to host the proposed institution. The recommendation was subsequently approved by the UNESCO General Conference in 1976, and, in 1981, an agreement establishing the legal basis for this autonomous international Centre was signed by the Government of Nepal and UNESCO. Finally, with the support of Hindu Kush-Himalayan countries and the Governments of Germany and Switzerland, ICIMOD was inaugurated in Kathmandu on 5 December 1983.

Mandate and Functions

The dual mandate for environment and development is reflected in Article 1 of ICIMOD's Statutes, which reads: 'The primary objective of the Centre shall be to help promote the development of an economically and environmentally sound mountain ecosystem and to improve the living standards of the mountain populations of the Hindu Kush-Himalayan Area which, for the purpose of these Statutes, includes Afghanistan, Bangladesh, Bhutan, China, India, Myanmar (Burma), Nepal, and Pakistan.'

The specific functions of ICIMOD, as set out in its Statutes, are to serve as:

- a *multidisciplinary documentation centre* on integrated mountain development, based on the systematic exchange of knowledge and experience through an organised information network;
- a focal point for the mobilisation, conduct, and coordination of *applied and problem-solving research activities*;
- a focal point for *training on integrated mountain development*, with special emphasis on the assessment of training needs and the development of relevant training materials for the training of trainers;
- a *consultative centre providing expert services* on mountain development and resource management to the HKH countries.

Development of Programmes

The Centre began operating in 1984, following the inaugural symposium of December 1983. While a diverse range of opinions was voiced on what ought to be done in mountain development, two main needs were consistently emphasised. The first was to arrive at a more practical understanding of the key socio-economic and biophysical linkages of development; the second was to mobilise professional and scientific expertise in a framework of regional cooperation.

The Centre's Phase 1 Work Programme was guided by three principal objectives:

i) to organise a series of state-of-the-art reviews of knowledge on key aspects of mountain development: Watershed Management, Rural Energy and Off-Farm Employment Generation;
ii) to develop working relationships with other regional and international institutions;
iii) to establish a Mountain Documentation Centre.

A number of international workshops and discussions were organised to discuss the reviews prepared and to identify programme priorities for the future.

After three initial years devoted primarily to institution-building and knowledge review, 1987 marked the beginning of a significant new phase of accelerated research, training, documentation and networking in the areas of Mountain Agriculture, Environmental Management, Infrastructure and Technology, and Population and Institutions. These activities were supported by the development of a Natural Resources Information System, Area Development Planning and Institutional Development (see Chart 1).

A quinquennial review of the Centre was completed in 1990. This made some important recommendations regarding the Centre's governance, organisation and programme thrusts. Based upon this review and the experience of the past five years, the Centre developed a Strategic Plan which identified overall priorities and the approaches to be adopted in future in order to fulfil its mandate. Following the approval of the Strategic Plan by the Board of Governors, the first Donors' Meeting was organised in Berne, Switzerland in June 1991. The purpose was to mobilise potential donors and solicit support for funding long-term programmes.

ICIMOD seeks to implement its Strategic Plan through a series of well-conceived and concrete medium-term plans, which are intended both to contribute to its long-term goals and to have a positive impact in the short and medium terms.

Highlights of 'ICIMOD Towards 2000: An Indicative Strategic Plan'

Goal   To promote ecologically sustainable development in the mountains and improve the socioeconomic well-being of their inhabitants.

Objectives To create awareness concerning the complexities and problems of proper use of mountain resources as well as to advocate the need for an integrated approach to sustainable development, and facilitate its implementation.

Strategy To overcome the twin problems of environmental degradation and poverty in mountain areas through applied and action research, training and demonstration, documentation and information exchange; as well as expert services, workshops, seminars, and conferences, in close collaboration with national institutions and professionals of the Region.

Clients Though the ultimate clients or beneficiaries of ICIMOD are the local mountain inhabitants, the primary clients are policy- and decision-makers, administrators, researchers, managers, professionals engaged in planning, implementation, and evaluation at middle or senior administrative levels.

Programme Formulation and Implementation

The process of programme formulation and implementation at ICIMOD may be described as being primarily consultative and collaborative. The process of consultation involves many participants, consisting mainly of professionals active in mountain development. Wherever feasible, representatives from government and NGOs have also been included. The main objective behind this consultative style is to arrive at a better understanding of the needs perceived to exist, actual priorities and what applied research and training is already being done.

In programme implementation, ICIMOD collaborates with national and local agencies, mainly for practical operational reasons. The cooperation of the various national agencies and local organisations in each country is essential for ICIMOD to organise activities at the national level.

ICIMOD's Experience

While ten years in the history of an institution is not a long period, it is nevertheless sufficient time for a Centre to establish its overall usefulness by demonstrating that it is using scarce resources efficiently to fulfil at least some of the initial expectations. Although ICIMOD's overall mandate was very broad ('to improve the mountain environment and economy and the well-being of mountain people'), it was clear from its statutory functions (documentation and exchange, problem-solving research, training and providing expert services) that it was not intended to be an aid agency, a university or a technology-generating centre (like other international agricultural centres). It was to operate rather as a development think-tank, playing an indirect, catalytic role in influencing decisions and action, through development of understanding, widening the knowledge base and enhancing awareness of mountain development problems and options. This clearly necessitated a close partnership with those who had a direct role in decisions and actions concerned with mountain development. Against this background, it is useful to document what ICIMOD has actually done so far.

Improving Understanding, Raising Awareness and Sensitisation

- Over ten different international and regional meetings have been organised, focusing on issues of relevance to mountain development, ranging from Energy and Watershed Management to Beekeeping, Horticulture and Institutional Strengthening for Environmental Management.
- About half a dozen national-level meetings have been organised, focusing on specific country-level issues. The main emphasis has been on mountain agriculture, employment generation, rural development, energy and participatory management of natural resources.
- ICIMOD has also collaborated with many international agencies to jointly organise meetings on topics such as Parks and People, Indigenous Knowledge, Snow Melt and Mountain Hydrology.
- ICIMOD has provided a forum for over one thousand professionals concerned with mountain development to present and exchange their views and assessments.
- About 180 documents (books, monographs and papers) on different aspects of mountain development have been published by ICIMOD so far.
- The Documentation Centre has a collection of about 12,000 documents on mountain development.
- ICIMOD is in regular touch with about 700 organisations in the HKH region through its distribution network.

Policy Reviews, Planning Methodologies and Studies

- Comparative reviews of mountain agricultural development experience in China, India, Nepal and Pakistan.
- Development of Mountain Perspective Framework.
- Geographical Information Systems (GIS) applications for resource assessments.
- Regional Integrated Economic and Environmental Development Planning for the Bagmati Zone, Nepal.
- Comparative review of off-farm employment generation policies and programmes.

Identification of Technology Options for Mountain Areas

- Use of Sea Buckthorn (Hippophea family) as a pioneer plant for restoring degraded lands in higher altitudes.
- Use of Urea-brick-molasses block as a supplement livestock feed.
- Use of polythene film technology in mountain areas.
- Promotion of sloping agricultural land technology for farming in unterraced sloping lands.
- Development of beekeeping using indigenous species.
- Preparation of user-friendly, low-external-input regenerative agricultural technology.
- Demonstration of different bioengineering approaches for rehabilitating degraded mountain areas.

Institutional Strengthening

- Comprehensive reviews of agricultural development agencies in the Hindu Kush-Himalayas.
- Promoting participatory development approaches, based on reviews of instances where forest resources are managed by the 'user group', i.e. the local population, and support for interaction and networking between NGOs and user groups.
- Providing training in areas such as Mountain Risk Engineering, CDS-ISIS (a bibliographic data base), GIS Applications, Participatory Rapid Appraisal Techniques, Tracer Hydrology and Energy Planning.
- Providing hardware support to selected agencies to conduct training in GIS Applications and Mountain Agriculture Planning.

Collaboration with International Agencies

- Watershed Management.
- Preparation of Chapter 13, UNCED *Agenda 21*.
- Technical Advisory Committee of the Himalaya EcoRehabilitation Project (India).

Thus, it is clear that ICIMOD's experience covers a wide range of development issues. Some of these have only been addressed once, while in a number of critical areas such as mountain farming, management of natural resources and GIS training, long-term programmes have been developed. The demand for ICIMOD's outputs has been growing substantially over the years, and while the impact of all the programme activities is not the same, certain ideas identified by ICIMOD are being integrated into national programmes. ICIMOD is now making efforts to consolidate its activities so that the breadth of coverage of the past will be replaced by more in-depth practical problem-solving work in selected areas.

Priorities for the Future

The Centre is assessing its activities in order to be more effective in meeting its mandate. In spite of numerous constraints, it is essential that scarce resources be put to the most effective use in areas where the Centre has a comparative advantage. The experience so far suggests the following priority thrusts for the Centre:

a) Reaching Mountain People

Although the Centre lacks the means to reach mountain people directly, its programmes must be of such a kind that their results can be readily translated into action. Working through its partners at national level, the Centre must focus on those problems where meaningful changes in the mountain economy and environment can be introduced as soon as possible. The work undertaken in the past has suggested the following broad directions which meet this criterion:

i) Improving local-level management of biomass resources through promotion of multipurpose trees, valuable grasses and other high-value crops, use of hedgerows, sloping agricultural land technologies, agroforestry schemes, low cost bioengineering methods, and different approaches to land rehabilitation and watershed management.

ii) Promoting participatory solutions to problems of environmental management.

iii) Identifying appropriate methods for training programmes and demonstrations, so as to improve dissemination of workable solutions.

b) Capacity Building

ICIMOD must seek ways to enhance the capacity of its partner organisations to plan, monitor, organise, manage and evaluate mountain development activities. It must work closely with selected agencies which have a similar mandate to its own, and identify concrete mechanisms through which better policies can be designed and more appropriate programmes developed and smoothly implemented. In order to promote this aim ICIMOD should focus on the following:

i)   providing appropriate hardware and software support;
ii) developing skills in areas critical for integrated mountain development;
iii) helping to develop information systems which can promote timely and appropriate decisions;
iv) encouraging greater understanding and sharing of experience between national agencies in different countries.

c) Preparing a 'State of the Mountains' Report

In order for ICIMOD to attract the national and international attention needed for integrated mountain development, it must now start to produce a regular assessment of mountain areas, problems and issues: a 'State of the Mountains' report. This should be based on reliable information on mountain areas, their resource conditions and important economic and environmental events. The project should start with national reviews, compiled in partnership with concerned agencies, which would be used to develop regional-level reviews.

d) Strengthening Regional and International Cooperation

ICIMOD's functions and resources do not permit it to be very diverse in its activities if it is to be effective. Consequently the Centre must find ways to channel the resources and activities of various regional and international agencies into mountain development problems. It should explore ways in which it can serve as a focal point for Chapter 13 of *Agenda 21* for the Hindu Kush-Himalayas.

Chapter 14

The Rural Non-Farm Sector in India

Issues of Relevance to Development in Central Asia

Thomas Fisher

Introduction

This chapter presents some salient issues from a study of the rural non-farm sector in India, conducted between 1992 and 1995.[1] Because of my limited knowledge of countries in Central Asia I have refrained from making comparisons with those countries, but I am confident that:

- readers will identify many similarities, as well as contrasts, with their own countries of origin or experience, from which they can draw their own conclusions;
- the rural non-farm sector is of great importance for sustainable development in any developing country;
- the recommendations of the study (Section 5) are of relevance to the goal of sustainable development, in that they suggest avenues in the search for sustainable employment for the rural poor, and analyse the institutions through which policies for sustainable development will have to implemented;
- the methodology used, for example in analysing economic subsectors, is of relevance to those who wish to promote sustainable economic development.

Objectives of the Study

The study of the rural non-farm sector, which was funded by Swiss Development Cooperation and the (Indian) National Bank for Agriculture and Rural Development (NABARD), aimed to develop policies for generating substantial productive and sustainable rural employment in India by the year 2000. India needs to generate 100 million jobs by the end of the century if it is to absorb new entrants into the labour force and wipe out the existing backlog of unemployment.

The primary focus of the study was thus very much on employment generation, rather than, for example, alleviating poverty or conserving the environment; but such issues cannot of course be neatly divided into compartments. One of the industries in India which employs the largest number of rural workers, outside of agriculture, is producing wood and wood products. Forest products are also an important livelihood for many of the poor, most notably the tribal peoples. However, forest reserves have been drastically depleted: the resource base for a crucial economic activity, especially for the poor, is being removed by environmental degradation. Likewise, rural-urban migration may lead to poverty and environmental pollution, but can be reduced by employment generation in the countryside.

Definition of Terms

We define the *rural non-farm sector* (RNFS) as comprising all economic activities in rural areas excluding agriculture and allied activities such as animal husbandry. Thus the sector encompasses, in the primary sector, mining and quarrying; in the secondary sector, construction, manufacturing, processing and repairs, in and outside of the household; and in the tertiary sector, trade, commerce, transport, communications and other services.

An economic *subsector* is defined by its final product (such as leather shoes and apparel), and includes all firms engaged in raw material supply, production and distribution of that product.

Various types of *institutions* influence the growth of the rural non-farm sector: regulatory bodies, promotional agencies, credit suppliers and representative associations such as trade unions.

Methodology

The non-farm sector study covered eight states in India, and in each state examined six to ten subsectors which generate substantial rural employment, or could potentially do so. We also investigated the institutions which influence the growth of the RNFS. Thus the study combines economic analysis with *institutional economics and analysis* (see Section 3). Methods used included literature surveys, analysis of census and other data and extensive field-work interviewing producers and representatives of institutions.

Profiles of each subsector were drawn up using the results of interviews, supplemented with secondary data. The subsectors were analysed using a framework based on the work of Professor Michael Porter of the Harvard Business School, in his book *The Competitive Advantage of Nations* (1990). For an introduction to the analytical framework, and an outline of the subsector profiles, see the Appendix at the end of this chapter.

As well as the subsector profiles, case studies were written of individual producers and institutions, and overall reports compiled for each state and for India as a whole. Local and national workshops were held to disseminate the findings.

Overview of the Rural Non-Farm Sector in India

Since independence India has focused its development efforts on agriculture and industrialisation. In the latter field, initiatives have largely aimed to promote urban industries, although there have been many attempts (largely unsuccessful) to attract urban-style industries into backward regions through a mixture of incentives, such as tax rebates. The rural non-farm sector has been neglected, even though it now provides some 40 to 50 million jobs (depending on the data source): between one sixth and one quarter of all rural employment. Moreover, this proportion is steadily increasing as non-agricultural employment in the countryside is growing faster (at 4.1% per annum, 1977–88) than agricultural employment (0.7%).

The main attention paid to the rural non-farm sector (RNFS) has been to protect and preserve a high level of employment in traditional subsectors like weaving and pottery, which are clubbed together under the term 'Khadi and Village Industries'. However, these schemes have mainly consisted of restricting production of certain goods to tiny village units, and of massive subsidies for production and rebates on sales. The result has been to confine the workers in these activities to low-technology, low-productivity and hence low-wage employment, rather than upgrading the industries to provide productive employment that could be sustained in the future.

Indeed, one of the primary reasons for neglecting the RNFS has been that policy-makers have failed to see it as a critical growth sector with the potential to meet national goals of employment and equitable income distribution.

According to data from the Census of India, 1981, which uses the National Industrial Classification (NIC), the following are the fifteen subsectors in the RNFS which generate the most rural employment:

NIC code	Description	% of Employment in RNFS
65	Retail Trade in Food, Beverages and Tobacco	8.8
92	Educational Services	8.1
90	Public Administration	7.2
96	Personal Services	6.4
50	Construction	5.8
27	Wood and Wood Products	5.3
70	Land Transport	5.2
23	Cotton Textiles	5.1
26	Textile Products	4.6
32	Non-Metal Mineral Products (e.g. pottery)	4.4
20–1	Food Products	4.1
22	Beverages, Tobacco and Tobacco Products	4.0
39	Repairs	3.4
99	Miscellaneous Services	2.8
69	Restaurants and Hotels	2.1

What is striking about these figures, and relevant to policy-makers concerned with employment generation, is that these fifteen subsectors account for more than three quarters of all employment in the rural non-farm sector. In other words, policy attention could be focused on a narrow range of activities.

The second striking point is the importance of services. Taken as a whole, trade, commerce, transport and other services account for 60% of all RNFS employment in India, which is comparable to the proportion in many other countries. And yet it has been a standard practice of policy-makers to focus on manufacturing and neglect services. Even within the secondary sector, the largest subsector in terms of employment in rural India is construction, again often neglected by policy-makers.

However, many of the subsectors listed above comprise traditional activities which have shown a decline in employment absorption and a low capacity to create new jobs. Thus we also used data on the *growth* of rural employment to identify the most important subsectors. The top ten high-growth sectors (1977–88) in descending order were: Manufacture of Electrical Equipment; Construction; Paper and Paper Products; Chemicals and Chemical Products; Recreational and Cultural Services; Land Transport; Real Estate, Business and Legal Services; Miscellaneous Services; Wholesale Trade in Food and

Textiles; and Manufacture of Jute, Hemp and other Fibre Products. Again, services and construction are prominent in this list. A number of modern manufacturing subsectors are also included, indicating their potential for generating rural (and not just urban) employment.

In addition, the study sought out *emergent* subsectors, which have developed too recently to appear in employment data as high-share or high-growth subsectors. These emergent subsectors were identified by informed experts, often in the field, from information on demand trends and the like.

It is important to recognise that many of these subsectors, apart from traditional artisanal subsectors and small service activities, generate wage employment, not self-employment. India has implemented massive programmes to generate self-employment among the rural poor: the Integrated Rural Development Programme is probably the largest project in the world to provide productive assets to the rural poor by extending them subsidised loans. While vast resources have been transferred to the poor, the programme has failed to generate substantial self-employment. In many cases the poor appear to prefer wage employment, if available, rather than having to bear the risks of self-employment.

Finally, businesses, or production units, in a particular subsector often cluster in one place. Each of these clusters is centred on a rural town or city which acts as a sub-regional trade centre. This clustering allows units to enjoy common infrastructure, sources of raw materials and marketing channels, to share information and to act together in addressing the constraints they face.

Importance of the Rural Non-Farm Sector

Full employment is just one of the potential benefits of increasing rural non-agricultural employment. There are many other benefits.

The RNFS provides higher wages than agricultural activities, particularly agricultural labour. The proportion of those below the poverty line in the RNFS in India is smaller than in the agricultural sector. The mobility of workers is also higher, allowing them more opportunity to seek better wages than they would have in agriculture.

The RNFS also demands diverse skills, and educational and skill levels in the RNFS are higher than that of the typical agricultural labourer. Indeed the sector contributes, at a mass level, to skill building among the rural population, as the vast majority of those who enter the sector each year are trained on the job.

Data for India indicate that the RNFS is less capital-intensive than urban industry, in terms of the ratio of capital to output. In terms of the capital-labour ratio, almost forty times as much capital is required to employ one worker in an urban factory as to employ one in a rural enterprise with six workers, of whom at least one is hired. As for energy use, over 70% of rural enterprises in India do not use power; most of these are of course small units generating employment for one self-employed person and perhaps some family members.

The rural non-farm sector is also more export-intensive than agriculture, and often even than urban industry. Among those subsectors which have a high share of employment in the RNFS in India are the ones which generate a large proportion of exports for India. These include Textiles and Textile Products, Gems and Jewellery, Handicrafts and Carpets. Of course, all these subsectors are linked to export units in large cities, but a significant part of the processing happens in rural areas.

Policies and Institutions

In looking at any sector of the economy it is important to understand the policy and institutional framework that influences it. The non-farm sector study focused on the following areas of policy: regulation; promotion; credit; and representation.

The experience of policy-makers is that well-formulated policies often get distorted by the time they reach the ground. Policies are implemented through institutions, which very often have their own goals, or get asked to implement conflicting policies. In such cases they implement policies selectively, or even reinterpret them to suit their own needs. Hence to analyse policies in a live context it is necessary to look at the implementing institutions.

For this purpose the study draws on various theoretical sources, including management studies, but above all institutional economics.[2] In institutional economics, the definition of institutions is broadened to 'the rules of the game in society, or, more formally, . . . the humanly devised constraints that shape human interaction'. These include:

> [the] set of rules, compliance procedures, and moral and ethical behavioural norms designed to constrain the behaviour of individuals in the interests of maximising the wealth or utility of principals. . . . A key distinction in this model is between agents

and principals. Put simply, agents work for principals, or . . . perform some service on their behalf which involves delegating some decision-making authority to the agent. . . . there is some latitude in decision-making power by the agent as a result of the inability of the principal to perfectly constrain the agent's behaviour. Most individuals are agents in one role as employees and principals in another role as consumers.[3]

Unfortunately it is not possible in this chapter to elaborate on the application of such a theory to the rural non-farm sector on the ground, although this has been attempted in the various reports produced as part of the study.

Regulatory Policies and Institutions

Regulatory policies affecting the RNFS in India include registration and incorporation; licensing for production and quotas for scarce raw materials; taxation; labour laws on wages, welfare and safety; environmental protection; consumer protection; and laws and rules for specific subsectors.

In fact economic activities in India are highly regulated, by a complex mass of often petty and even contradictory rules. The impact on the rural non-farm sector has been mixed. On the one hand, the small and widely dispersed rural units may be beyond the reach of regulatory authorities. On the other hand, the very industrial structure of subsectors may be determined by the regulatory framework. Thus units which employ less than twenty workers (without power) or ten workers (with power) are exempted from the Factories Act, which contains many of the labour provisions. As a result many larger units simply fragment, with multiple units below the specified size being independently registered, even though the units are next to each other and are clearly part of the same production process. Protection of traditional activities, as well as restrictions on imports of modern technologies, have also led to the technological backwardness of numerous activities, and hence their low productivity and low wages.

The regulatory framework has been particularly influential in those cases where the state allocates public goods or resources. In such cases regulations tend to favour large players and have led to extensive collusion and corruption between them and the regulators. In such cases the powerful are able to monopolise licences, buying them at less than their full market value, in order to extract the maximum rent possible from the resources secured through the licence. To give a concrete example, the regulations about granting leases for mining 'minor' minerals in India have ensured that only those with spare capital and who were in collusion with the regulatory officials could obtain the leases, whether they had any expertise in mining or not. Moreover, leases were awarded for very small plots of land and very short periods of time. As a result, standards of mining have been very poor, with almost no proper technology used, leading to appalling conditions for labourers and substantial environmental damage, as lease-holders have sought to extract as much mineral as possible as quickly as possible through 'slaughter mining'.

The sheer complexity of the regulations and the power of the bureaucracy in India have led to substantial harassment by officials, often leading to bribery and other corruption. However, it is not just a matter of extra-legal payments. Complying with regulations involves transaction costs for producers, which become more significant the smaller the scale of operations and the more bureaucratic the regulatory institution. Thus the chief complaint among producers interviewed in the study was not extra-legal payments (which could be regarded as simple business costs) but the fact that they faced a multiplicity of regulatory institutions, many of which had several representatives each, with whom the producer had to deal. As one producer put it, he goes by the rule of thumb of 'one inspector a working day'. To take another example, in 1986 the government introduced an 'Environmental (Prevention and Control of Pollution) Act', coupled with specific acts for water and air. These laws are implemented by state and national Pollution Control Boards. However, implementation is very erratic, and in many cases officials use the laws for monetary gain by refusing clearance to units to start operations or threatening prosecution of existing units.

On the reverse side, many of the institutions entrusted with regulatory functions maintain that they do not have the reach or powers to implement the laws effectively. To continue with our example, the Pollution Control Boards maintain that they do not have adequate powers to prosecute for non-compliance, and that they can only withhold initial clearance.

Promotional Policies and Programmes

India has employed an extensive list of development policies to protect traditional rural industries, to promote industries in backward regions, to promote self-employment among the rural poor, and to encourage specific subsectors (such as textiles, leather, food processing and silk production) through fiscal incentives and the creation of subsectoral promotional institutions. Policies have included developing infrastructure, establishing backward linkages (links between producers and raw materials suppliers), upgrading technology and skills, promoting forward linkages (marketing), and providing subsidised loans.

Some of the drawbacks of these policies have been:

Protection rather than development: India has sought to protect employment in traditional subsectors rather than upgrade them to provide productive and sustainable employment, albeit for a smaller number of workers, in the future.

Intervention rather than facilitation: The state and central governments have often intervened in economic activities by establishing numerous state corporations and institutions which play a direct role in the subsector, for example by undertaking production and marketing themselves. There have been far fewer policies to facilitate growth of a subsector, for example through provision of an appropriate regulatory and promotional framework.

Inadequate attention to promoting links between RNFS activities and the primary resource base: Since agriculture is one of the main sources of inputs as well as of demand for the RNFS, growth in agricultural output and productivity is important for the growth of the RNFS. Yet little policy attention has been paid to such links.

Inadequate attention to physical infrastructure: The state and central governments in India have failed to provide the necessary infrastrucure, such as roads, communications and power, even for booming subsectors. Instead vast resources have been wasted trying to attract industries to backward areas where they do not want to be.

Inadequate attention to social infrastructure: India has not ensured universal literacy, let alone universal primary education, which severely limits the possibility of skill building within the population.

Inadequate focus: India has not focused promotional efforts on the most important subsectors which have potential for employment generation in the future. Instead, resources have been wasted in promoting other subsectors, often for political reasons, even when they have little potential for generating productive and sustainable employment. As a result many important subsectors have been neglected, particularly among the services.

To turn to analysis of the institutions implementing these promotional policies,[4] most are initiated and owned by government. As a result they have had very little autonomy: the Chief Executive Officers have been at the whims and fancies of the political leaders controlling the government departments under which their institutions come. To make matters worse, the institutions are evaluated by non-commercial measures of performance and rely on the state exchequer for their finances. Since they are not dependent for their survival on service fees from industry, funds have often been allocated to projects for which RNFS producers themselves felt no need.

Secondly, most of these institutions have been given a multiplicity of goals which are difficult to translate into specific objectives and methods, except in the form of fairly rigid programmes. Those institutions which were set up to provide technological inputs for specific subsectors have been relatively more successful.

Combining regulatory and promotional functions, as in many of the government departments and directorates, has proved particularly dysfunctional. Since promotion requires greater creativity and effort, it proves easier for such agencies to concentrate on regulation, which over time comes to overshadow promotion completely. This is not hard to understand. Functionaries can exercise greater power in regulating and releasing or withholding funds. There is little incentive to perform the promotional role, particularly because there is no system to recognise and reward good work.

The crux of the problem can be understood in terms of the principal-agent relationship. In the current institutional framework for promotion in India the state is the principal which appoints the promotional agents to act on its behalf. The situation can be remedied by establishing a direct relationship between the producers and the promotional officials, in which the producers, who are the ultimate users of the services, are the principals and the promotional officials their agents. Such a relationship would mean not only that the promotional agents were more accountable to the producers, but would also lead to a better service, since producers would bring to the relationship a deeper knowledge and understanding of their promotional needs than the state has. Some of the most successful promotional institutions in India have thus been the Export Promotion Councils, which were established by the government, but in collaboration with exporters, who are significantly represented on the Boards of the Councils. The Councils depend for their funding not just on government budgets, but also on fees for services provided to exporters. Finally, they have not taken up the marketing function themselves as government-owned corporations, but facilitate exports by arranging exhibitions and meetings between buyers and sellers, providing information to producers on export markets, advertising abroad on behalf of industry, and so on.

Thus the state need not necessarily be responsible for managing promotional activities, but could merely encourage private parties to do so. The provision of common services, such as gathering market information, marketing the products of small businesses, purchasing raw materials in bulk, providing quality testing facilities, can often be organised more effectively through producer associations, especially of active producers in dynamic clusters, than by state-owned corporations.

Credit Policies

Among the primary objectives of credit policies[5] in India since independence have been efforts to extend credit into rural areas and to replace informal credit systems (lending by landlords, money-lenders, traders, etc.), which are often still regarded as exploitative. The commercial banks were nationalised in 1969 and forced to open branches in all rural areas of the country (in many areas there is a branch for every 15,000 people). The banks have been given targets for lending to specific priority activities and specific groups (such as those below the poverty line, tribal peoples and those belonging to the lower castes), which has been heavily subsidised. Altogether this has amounted to one of the largest poverty alleviation programmes ever attempted.

However, if we judge these initiatives by their results, they are disappointing. Below market interest rates have resulted in excess demand and credit rationing. Political influence in credit allocation is high. And for many of the rural poor, getting loans from the commercial banks is a time-consuming and even harassing experience, all adding up to high transaction costs. These costs include out-of-pocket expenses, wages foregone because of the time spent obtaining the loan, bribes to bank officials, and deliberate over-charging, in collusion with the bankers, by the sellers of assets which are given to the borrower in kind. A World Bank study of credit in India in 1994 (see note 5) indicates that the effective rates of interest (including transaction costs, but excluding the subsidies) can be as high as 26–33%, even though the interest rate paid to the bank is only 12%;

while for small loans, transaction costs amount, on average, to 40% of the loan amount.

The effective interest rates are still lower than those in the informal sector, where rates range from 40% to over 80%, although the transaction costs for such loans are almost zero. In addition to low transaction costs, informal channels of credit, which usually lend only locally, enjoy other advantages over the commercial banks: their accessibility, the minimal time taken to process a request, and the greater ease with which they can screen borrowers, give them incentives to use their credit for the stated purpose (for example by linking credit for production with the supply of an input), and enforce repayment. Thus the share of credit provided by informal sources continues to be high in many rural areas, in spite of policy initiatives to displace them.

Moreover, because bank officials have not been given discretion about lending to the poor at below market rates, they have become cynical about it, believing that it cannot be profitable[6] and requires too much follow-up, which in the absence of the necessary staff is not possible, that the projects are mostly not viable, and that the recovery rates are low because of the perception by beneficiaries that such loans are essentially grants. Repayment rates have indeed been appalling, because of inadequate appraisal of loans, unrealistic targets for direct lending, misuse of funds, and not least, mass-scale loan waivers during elections.

Representational Policies and Institutions

By this phrase we mean the set of policies which ensure participation by producers, workers and consumers in the process of formulating, implementing and monitoring policies and regulations that affect them, and which give them control over promotional agencies meant to serve them. There are indeed many trade and producer associations, societies, cooperatives and trade unions which lobby for the interests of their members. However, there are no effective policies in the country to ensure representation of producers, workers or consumers in policy-making institutions. Policy-making is considered the prerogative of the bureaucrat, with policy directions coming from the political party in power. Those affected by the policies can only lobby politicians and bureaucrats from outside the decision-making process.

In spite of such a policy scenario, in the course of the field-work for the study we came across many producer associations, both formal and informal, and many of them unknown to officials. The most successful of these are the associations of producer-entrepreneurs in a specific local cluster of enterprises who have come together to tackle common problems faced by the units in the area. In two places we found such associations which were establishing very large industrial estates, one for ready-made garments and the second for diamond-processing. These estates included independent water and electricity supply, effluent treatment plants, an exchange for trading diamonds, export clearance facilities, and so on. Both were established with some government support, and both indicate the potential of such associations to establish and manage facilities for their subsectors.

Unions in India are very strong in formal, often public sector enterprises. However, their presence in the mainly informal rural non-farm sector is very weak. Consumer associations are still at a nascent stage.

Recommendations

Preliminary

The rural non-farm sector needs to be given adequate policy attention as a sector in its own right, in addition to agriculture and (largely urban) industry. The RNFS not only has the potential to generate substantial rural employment, but also to bring many other benefits: increasing incomes and building up the skill base of the rural population; lower capital intensity; lower energy requirements; and a higher contribution to exports. The RNFS can thus provide productive and sustainable employment to the rural population and reduce rural-urban migration.

To develop the RNFS, a government needs to implement certain general development policies, which act as a stimulant, indeed which are a precondition for the growth of the RNFS. These include achieving dynamic agricultural growth and improving physical and social infrastructure.

A substantial part of employment in the RNFS is in the processing of primary products. Thus one of the greatest stimulants for the growth of the non-agricultural economy in rural areas is dynamic agricultural growth. In addition to providing inputs for processing, agricultural growth creates demand for inputs and services from the local economy, and generates income surpluses, which generate demand for all kinds of consumer goods and services. Apart from agriculture, other primary sector activities that can serve as an important basis for RNFS livelihoods are livestock-rearing, forestry and mining. To the extent possible, policies should encourage rural people to add value by processing these primary products, rather than merely trading them.

Both physical and social infrastructure are necessary for the development of the RNFS. The sector requires better-educated workers, so that universal primary education is a vital first step for providing such a workforce. However, it is only a first step, as becoming a productive worker in the RNFS requires the acquisition of skills, for which vocational training is desirable. Providing basic physical infrastructure to each village (for example roads, communications, power and water) would also give a boost to the RNFS. However, in most cases the resources do not exist to provide every village with such infrastructure, and appropriate technologies, for example to provide power locally, need to be installed. More intensive infrastructure investment should be focused on those areas where the returns on investment are higher (for example dynamic clusters in particular industries).

Focused Promotional Activities

Choice of Subsectors: The last point in the previous section draws attention to the need to focus development activities. The rural non-farm sector has enormous diversity, ranging across the primary, secondary and tertiary sectors, and encompassing tiny units all the way up to large factories. The first policy choice that needs to be made, therefore, is which subsectors to select for policy attention, which is why the study analysed individual subsectors.

Sustained growth in demand, both domestic and for export, is the single most important factor determining the growth of a subsector. Seeking to promote subsectors without this sustained growth in demand has been a disaster in India. For example, in spite of vast subsidies, promotional activities and regulatory restrictions, protection of traditional employment in hand-woven cloth has failed to generate sustainable employment for the weavers concerned and has most likely damaged the growth of the subsector.

From the analysis of the RNFS in India, subsectors could be selected for policy attention from the following categories:

Emergent subsectors which at present support relatively little employment but which in the opinion of informed persons have a potential for growth in the near future because of unmet demand. Because of their emergent status, selected subsectors under this category may initially require substantial promotional inputs.

High-growth subsectors which have registered a high growth rate of demand in the domestic and/or export markets. The role of policy here would be to ensure that the favourable conditions created by growing demand for the products of such subsectors are backed up by adequate availability of raw materials, infrastructural support, training, product and process upgrading services, credit, and marketing services.

High-share subsectors which at present account for a high share of rural employment, even though their demand prospects are uncertain. The role of policy here would be to ensure that the existing levels of employment are protected, but with minimal distortion through subsidies, product reservations or production controls. Where the demand conditions are unfavourable and the subsector faces declining prospects, another role of policy could be to ensure substantial retraining of workers and their redeployment.

Comprehensive Approach

Once subsectors have been selected, the perspective taken on them must be comprehensive. Often promotion of rural industries has simply had the objective of promoting individual manufacturing units. Even in cases where specific subsectors have been identified, this often translates merely into preferential allotment of land, power, credit and so on to units in those subsectors. What is needed for the subsector to thrive is the emergence of enterprises and other entities (for example, producer associations and research institutes) which provide a variety of services to the units in the subsector. These services include product design, technical development, training, packaging, market research, advertising, sales promotion, and so on.

Thus there is a need to review a subsector as a whole and to identify the missing links, or ruling constraints. Generally a subsector would not be seen as promising unless the demand conditions are favourable or at least potentially so. In the latter case, some work may have to be done to relieve the constraints on demand. Apart from that, the ruling constraints will be found in factor conditions,[7] related and supporting industries and the structure of the industry. Such constraints need to be addressed systematically, some through promotional measures and others by developing or modifying regulation. In general it is the recommendation of the study that, except in the case of emerging subsectors, government should play only a catalytic role, encouraging initiatives by producer associations and industry-sponsored institutions.

One last point of caution needs to be noted: it takes a long time to promote a subsector, and the perspective adopted should not be less than a decade. Policy-makers should not be impatient in this period and should avoid the temptation for the state to 'do it itself'.

Choice of Location

Once subsectors have been selected, it is also important not to waste promotional resources by seeking to assist all units. Subsectors tend to develop around geographical clusters and these can be used as a focus for promotional inputs. The emergence of a cluster may be a function of nearness to raw materials, the existence of skilled labour or simply a matter of chance. But once the cluster begins to emerge, government policy can facilitate its orderly growth. Attempts at forcing the creation of clusters which are not based on any natural advantage do not usually prove successful.

Concentrating policy efforts on existing clusters has two advantages. First, since most infrastructure is capital-intensive and indivisible, having a large number of users in a geographical concentration reduces costs per unit and makes for quicker returns on investment. Second, clustering also makes it possible to provide easy and quick access to related services such as credit, product design facilities, training centres, pollution control, etc.

If any attempt is considered necessary to create clusters in less developed regions, it is best to invest in specialised infrastructure, such as a research institute or a training centre, which can act as magnets for drawing private capital. This form of intervention is better than providing subsidies to firms to locate facilities in a region where they would otherwise prefer not to.

It should be noted that the choice of locations, that is clusters, for attention, should be sequential, and not parallel, to the choice of subsectors. Thus only clusters in selected subsectors should be targeted for promotional effort. This is suggested to ensure that the limited amount of promotional energy available is used for the most promising clusters of the most promising subsectors (from the point of view of employment).

Choice of Unit Size

The third policy option that needs to be exercised is a decision on which size of unit should be given maximum incentives. The traditional policy approach for the RNFS in India has been to promote cottage-scale units: the individual weaver, the potter, the village shoe-maker. Even when government has promoted small-scale industries, these have employed on average only five workers. In short, RNFS promotion has been confined to very small manufacturing units. This has meant that scarce promotional resources have been wasted, as it is very difficult and expensive to reach many cottage or tiny units which are often widely dispersed. The conclusion drawn from the study is that promotional efforts should be focused on somewhat larger units, for the following reasons: these are units which are easier to identify; they already have a core of entrepreneurship and capital on which further dynamic growth can be based; and they are better able to assimilate and effectively use promotional inputs. Moreover, they provide opportunities for wage employment, thus reducing the pressure to be self-employed: not all rural workers have the capacity to be self-employed.

However, the general principle should be to establish units of whatever scale is necessary to capture as much of the value addition as possible. If the ruling constraint in a livestock subsector is facilities for processing, these should be promoted at whatever scale is most appropriate. Even if this means an urban processing unit, the demand generated for the livestock products will generate substantial rural employment.

Reorienting the Institutional Framework

Regulation and Regulatory Institutions: In promoting any subsector it is useful to review all the laws that apply to it and examine to what extent these need to be supplemented, amended or curtailed in order to encourage its growth. It is obvious that regulations in general are needed and cannot be abolished. However, the experience of rural India indicates that it is better to draw up regulations which are simple and intelligible and then to enforce them, than to develop a mass of complex regulations for an ideal situation, most of which do not then get implemented. However, some regulations, for example relating to labour and environmental protection, can prove helpful even when not fully adhered to, because they give a handle to workers and consumers with which to fight, for example in the courts, against infringements of these laws.

It should be mandatory to consult producers or representatives of producer organisations in all matters of policy-making relating to their subsector. There should be adequate transparency and attempts by the state to increase awareness of regulations through the media and publicity. The names of bidders, licensees, suppliers, etc. should be published in newspapers or commercial gazettes so that those who have been awarded government contracts or licenses, and the prices they paid, are known. This would stimulate competition in future rounds and enhance the accountability of government officials. Regulatory departments such as excise, sales tax, directorates of industry, inspectors of factories, etc. should periodically hold open fora where producers are invited to air their grievances and officials also get a chance to educate the producers. In order to ensure that lower-level field officials do not harass producers, an annual rating of such officials by the producers under their purview should be made a part of the appraisal process. Such feedback could be sought through the postal system. Limitations on the authority of inspectors to inspect units may also be necessary. Some such are now being introduced in various states in India, for example the rule that factory inspectors may inspect only 5%, chosen at random, of all small-scale units, or that an inspector must receive written permission from his superior before inspecting a unit.

Finally, while some level of regulation will always be needed, it does not follow that it must be done by state functionaries. There has been little thought in India of alternative regulators, but there are at least four other possible ways of regulation:

- regulation through third-party professionals, such as auditors, chartered accountants, company secretaries and cost accountants;
- self-regulation by producers' associations;
- regulation of units by worker trade unions;
- regulation of producers by consumer and citizen groups.

The first can be used for monitoring compliance with regulations; the second, for vetting producers for credit and monitoring repayment performance. The third may enforce workers' rights to better wages and safe working conditions; the fourth could serve to ensure product quality, safety and fair pricing, as well as compliance with environmental laws.

Promotional Institutions: These should be:

- subsector-specific, so that they attain a certain depth in understanding the problems of the subsector and identifying ruling constraints;
- cluster- or region-specific, so that they identify fully with local constraints;
- representative of the stakeholders involved – in particular, the producers they are meant to assist should be represented on the board;
- autonomous of the government;
- focused – they need to identify the key constraints facing a subsector and develop a clear strategy of intervention;
- economically viable – other than the initial start-up funds to which government may contribute, they must be capable of sustaining themselves by generating their own resources.

An exception to the last suggestion could be made for emergent subsectors; the state should be more deeply involved in promoting them, since pilot projects may be needed to establish the technical and economic feasibility of the activity. Even with promotional institutions established by the state, however, the primary role should be to foster private enterprise and not compete with it. Thus state-owned corporations have often established their own production facilities, often poorly managed and under-utilised, which seek to compete with private entrepreneurs, instead of creating supporting facilities so that such entrepreneurs can establish or develop successful businesses.

Promotional institutions sponsored by the state should also be opened to competition. The most effective means to do this is to reduce their budgetary allocations and increase their dependence on earning fees for services provided to producers or other clients. Only this can ensure that the organisations constantly strive to do relevant work and are also responsible for marketing their services among the producers they are supposed to assist.

Credit Institutions: These should be flexible in their lending policies. First, they should be allowed to set their interest rates so that they can recover their actual transaction costs in making loans to the poor; otherwise, they will always have a disincentive to undertake such lending. In fact bankers should be encouraged to see lending to the poor as a potentially profitable activity. In India targets for subsidised lending, coupled with political interference, have removed all discretion from bankers in lending to the poor, and they have become cynical and treat every poor applicant as a wilful defaulter to whom they would never lend if they did not have to meet the relevant targets. The government in its turn should of course avoid any loan waivers which suggest to borrowers that they may not have to repay loans, which only confirms the bankers' perception of poor borrowers, and has had disastrous consequences on repayment rates in India.

On the bankers' side, simple and easy procedures must be devised for lending to the poor or to small rural producers. The vast majority of loans required by the poor are in fact consumption loans, for example for medical expenses and emergencies, marriage ceremonies, or for smoothing food consumption over different seasons. The banks should thus not restrict themselves only to loans for productive economic activities. Initially small loans can be granted, and when these are repaid regularly ever larger loans can be sanctioned, thereby giving a powerful incentive to repay. In the case of production loans procedures should be sufficiently flexible. For example, for small loans no distinction should be made between fixed term loans for purchasing assets and working capital loans. In India loan amounts have often been fixed for each particular productive activity. This has had very poor results as bankers are not able to determine loan amounts according to local circumstances or respond to the specific needs of the entrepreneur. Loan repayments should follow a seasonal pattern appropriate to the business cycle of the particular subsector, rather than a rigid monthly schedule.

Bankers are often naturally cautious, and they must be given incentives to take the risk of lending to the poor or to small producers. In India bankers are evaluated according to deposit mobilisation, branch housekeeping and meeting targets set under the various poverty alleviation programmes. They do not get good marks, and in fact may get into trouble, if they make innovative loans or promote off-beat activities.

Because of the diversity of the rural non-farm sector, a common complaint against bankers in India is that they do not understand the business of the applicant and are thus extra-cautious. This could be overcome by establishing bank branches in industrial clusters which specialise in lending to the particular industry in that cluster, and can thus build up the necessary expertise. This would also reduce the time required for bankers to assess applications, as they would know the business, and in many cases the entrepreneurs as well, from previous lending transactions. If specialised branches cannot be established, then at least a technical cell in the regional headquarters is required, which can advise branch managers on technical aspects of the loans under consideration.

However, reaching a large number of small producers is often difficult, time-consuming and expensive. The banks should devise innovative ways of extending credit to such producers, particularly through intermediary structures or distribution channels. In India self-help groups are growing in number. These are small groups of individuals, known to each other, who come together periodically to pool very small savings. Once the pool becomes large enough, the group starts lending to its members, taking its own decisions about the size of loans, interest rates and repayment schedules. Banks can link up with these self-help groups, providing bulk loans for the groups to lend on to their members. The group structure usually ensures high repayment rates, while assistance to strengthen the groups may be provided through NGOs as well.

A second distribution channel is to extend credit to traders who supply raw materials and job work to rural producers. In many subsectors in India, including most handicrafts, it is such traders who have sustained much employment, through a putting-out system in which they supply the raw materials and designs and pay a piece rate to the producer when the finished good is returned. For a long time in India such linkages have been very popular with the poor producers themselves as they provide employment without the risks of self-employment, such as raising capital to buy raw materials or developing designs to cater to distant markets. However, if credit is extended to traders for on-lending to small rural producers it is vital that this be done to a sufficiently large number of traders in the same area, to ensure competition among them and thus that their margins are reasonable. Credit should be extended to traders for stocking raw materials as well as finished goods. The advantage of this condition is that it is easier for bankers to operate and more secure, since they can insist that stocks are pledged as collateral for the loan.

Representational Institutions: These should be encouraged. Producer associations can perform many functions, such as purchasing raw materials in bulk, providing technical assistance to their members, undertaking joint market promotion strategies, managing common infrastructure, and even collecting taxes from their members on behalf of the government. To encourage such institutions is not an easy task, and it may take a long time to mobilise them, especially among the poorer producers. This is not a role that governments do very well, and should be left to non-governmental organisations to the extent possible within the prevailing political framework. However, the government can assist by devising appropriate means of incorporation (in India there are societies, trusts, cooperatives and non-profit companies in addition to partnerships and corporations). One way to strengthen producer associations is to devolve resources to them, which could attract strong leadership. Such resources might come from an additional levy on excise duty, or by allowing the associations to collect taxes and keep a small percentage of the revenue.

In addition to producer associations, workers need to form unions to fight for their rights, such as wages and benefits, and consumers to form consumer associations which can take up issues of consumer protection. A major step has recently been taken in India through the formation of consumer courts in each district, where cases can be cleared more swiftly than through the formal legal system. This should encourage more activism on the part of consumers against faulty and dangerous goods.

Alongside the promotion of representative institutions, a general policy needs to be put in place, whereby producers are represented on the boards and committees of promotional institutions and policy-making bodies which directly affect them. Policies and programmes devised without any genuine consultation with those who are to benefit from them have rarely proved successful.

Appendix: Methods Used in Analysing Subsectors

Competitive Advantage versus Comparative Advantage

To understand what factors determine whether specific subsectors will be able to grow in the future, we use the concept of 'competitive advantage', based on the work of Professor Michael Porter of the Harvard Business School. Porter's research study of ten nations (USA, Britain, Germany, Sweden, Switzerland, Japan, Italy, Denmark, Korea and Singapore), published in *The Competitive Advantage of Nations* (Porter, 1990), presents a useful model for understanding which industries of a nation's economy can become competitive, and how. The shift in paradigm from the classical economic theory of the *comparative advantage of nations* to *competitive advantage* is fundamental to our way of thinking about how economic growth is fuelled. The traditional wisdom of focusing on those industries in which a country occupies a cost advantage is no longer valid since, according to Porter, other countries will innovate to increase their productivity and lower their costs.

Determinants of a Subsector's Success

Porter identified four determinants shaping the environment in which firms compete, which may promote or impede the creation of competitive advantage:

1 *Factor conditions*. The nation's position in factors of production, such as skilled labour or infrastructure, necessary to compete in a given industry.
2 *Demand conditions*. The nature of home demand for the industry's product or service.
3 *Related and supporting industries*. The presence or absence in the nation of supplier industries and related industries [including services] that are internationally competitive.
4 *Firm strategy, structure, and rivalry*. The conditions in the nation governing how companies are created, organised, and managed, and the nature of domestic rivalry.[8]

These four determinants are presented by Porter in the form of a 'diamond'. The diamond is a mutually reinforcing system. Advantages in one determinant can create or enhance advantage in others.

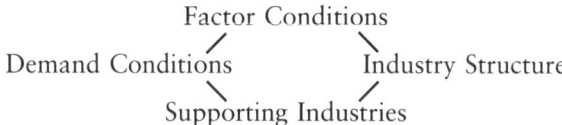

Drawing on this theory, and to identify the ruling constraints on the growth of each subsector, the non-farm sector study gave scores ranging from 1 (highly unfavourable) to 5 (very favourable) to each subsector under each of the following headings:

Demand Conditions
• Size of domestic market
• Large number of buyers
• Sophistication of buyers
• Presence of multinational buyers
• Growth rate of domestic demand

Factor Conditions
- Physical raw material availability
- Human availability and skill
- Knowledge and research capability
- Capital availability
- Availability of infrastructure

Industry Structure
- Large number of firms (no monopoly)
- Efficient size of firms
- Existence of rivalry among firms
- Formation of new firms (no barriers to entry)
- Congruence between sector and firm goals

Related and Supporting Industries
- Design and product development
- Marketing, market research and advertising
- Training
- Research and development
- Component and machinery suppliers
- Existence and effectiveness of producer associations
- Existence and quality of promotional institutions

Porter found that the *most* critical determinants of a subsector's success are:

(i) the existence of a large and discerning domestic market;
(ii) the existence of a large number of firms with intense competitive rivalry among them.

Intense competition leads to the creation of supporting industries and services which further enhance the innovation and competitive advantage of the subsector. It follows from this that factor conditions, which in classical economic theory were considered the basis for comparative advantage, are not the most critical determinant. Porter found that in many examples he studied, traditional factor conditions (such as raw materials) were in fact adverse to begin with (e.g. in the steel industry in Japan and ship-building in Korea). Basic raw materials were often imported, and other factors such as skilled manpower, technology, specialised infrastructure, design and marketing skills were created by firms either singly or through trade associations, under pressure not to fall behind.

The Role of Government

An important point to note in Porter's model is that government does not appear as a determinant in its own right. Instead,

The proper role for government policy toward a nation's industry is to stimulate . . . dynamism and upgrading . . . [and] is to unleash and even amplify the forces within the 'diamond'. This creates opportunities, and pressures, for continued innovation. . . . By stimulating early demand, confronting industries with the need for frontier technology through symbolic cooperative projects, establishing prizes to highlight and reward quality, encouraging rivalry, and other policies, the pace of innovation and upgrading is accelerated. [9]

The government can influence each of the four determinants either positively or negatively:

Demand conditions can be shaped by government, though its role is a subtle one. State bodies establish local product standards and regulations that influence product markets. The state is also a major buyer of many products. Its insistence on quality and high performance can thus influence the industry.

Factor conditions are affected by subsidies, trade policy, policies towards capital markets, education, infrastructure (roads, telecommunications, ports, services) and so forth.

Industry structure: governments have considerable influence on the ways firms are created, organised and managed, and how they compete. Competitive advantage requires that companies take a global approach to strategy. Government policy plays a role in this process, through regulations on foreign direct investment, exchange and import controls and other mechanisms. The goals of individuals and firms, though often shaped by social attitudes and structures, can nonetheless be influenced by the state, particularly through taxation, anti-trust laws, laws governing labour markets (and thereby labour productivity), the regulation and protection of domestic industry, and so forth.

Related and supporting industries can also be affected by government policy. Porter found, for example, that an unconstrained and innovative media – TV, radio, magazines, newspapers, and so on – were sources of competitive advantage in marketing and in reducing information and communication costs. The state can also reinforce and nurture the formation of geographical 'clusters', for example through the provision of concentrated infrastructure facilities, which are beneficial for building competitive advantage.

However, 'Government policy toward industry must recognise that the 'diamond' is a system, which makes policies in many areas interdependent. The weakest link constrains the development of an economy, so that progress is needed on each determinant.'[10]

Structure of the Subsector Profiles Compiled for the RNFS Study

1 Overview
2 Structure and Dynamics
   2.1 Employment, Output and Value Added
   2.2 Structure of the Industry
3 Product Mix, Demand Trends and Market Channels
4 Raw Materials, Infrastructure and Supporting Industries
   4.1 Raw Materials
   4.2 Infrastructure
   4.3 Supporting Industries
5 Technology, Skills and Training
6 Financial Needs and Sources
7 Policies and Institutions
   7.1 Regulatory
   7.2 Promotional
   7.3 Producer Organisations
8 Prospects and Constraints
   8.1 Demand Conditions
   8.2 Factor Conditions
   8.3 Industry Structure
   8.4 Related and Supporting Industries
9 Recommendations
   9.1 Demand Conditions
   9.2 Factor Conditions
   9.3 Industry Structure
   9.4 Related and Supporting Industries

Notes

1 For greater detail see Thomas Fisher, Vijay Mahajan and Ashok Singha, *The Forgotten Sector: Employment and Enterprises in Rural India* (Oxford and IBH, Delhi and Intermediate Technology, London, 1997).

2 For readers who have little respect for economists, especially neoclassical economists, the developing discipline of institutional economics will come as a surprise and may, hopefully, restore their confidence in the value of economics as a social science. Institutional economics is also becoming more accepted in the mainstream of economics, especially since one of its leading proponents, Professor Douglass North, was awarded the Nobel Prize for Economics. His book *Institutions, Institutional Change and Economic Performance* (Cambridge University Press, 1990) is recommended as a good introduction for non-economists as well as economists.

3 Douglass North, *Structure and Change in Economic History*, Norton, New York, 1981, pp. 201–2.

4 For this particular analysis we draw on a very useful book by Arturo Israel, *Institutional Development: Incentives to Performance* (World Bank, 1987).

5 This section also draws on a more detailed study in India: Price Waterhouse, *Financial Services for the Rural Poor and Women in India: Access and Sustainability. Interim Report* (World Bank, 1994), in which my colleague, Vijay Mahajan, was directly involved.

6 According to the World Bank credit study, the annual cost to the bank of extending a small loan amounts to 24–28% of the outstanding loan amount.

7 For an explanation of this term, see 'Determinants of a Subsector's Success', in the Appendix.

8 Porter, 1990, p. 71.

9 *Ibid.*, pp. 618–20.

10 *Ibid.*, p. 682.

Index

Achaemenid 13, 17, 33, 40, 49
Afghanistan 5, 20, 27, 30–40, 50, 56, 172, 200–1
Agenda 21 119, 207
agriculture 10, 50, 73, 81, 105, 109, 114, 126–27, 156, 161–63, 168, 180–84, 192, 197–98, 205–20
alienation 108
Almaty 162–66
Altai 9, 21, 23, 126
Amu Darya 10–23, 28, 33, 35, 42, 52, 162
Amur 4, 9, 26
animal husbandry 10, 44, 54, 105–16, 145, 209
Arabic script 42–43
Arabs 18–19, 36–37, 110
Aral Sea 18–21, 32, 52, 85, 161–66
Aramaic 40–42

Bactria 13, 35, 40, 45
Bactrian 41
Baikonur 53, 160
Baltic Sea 8
Baluchistan 11, 16, 27, 172
Bhutan 10, 79
biomass 126–39, 140–45, 206
Bon religion 32, 36
Buddhism 35–37, 39, 46
Bukhara 19–23, 27, 29, 38, 42, 48
Bukharan Jews 35
Buryatia 52
Byzantium 12, 18

camel 9, 48, 96, 113
capitalist 75, 77, 80
carrying capacity 9, 51, 96, 101–2, 193
Caspian Sea 5–12, 22, 26, 55, 159, 182

Central Asia, definition 4–8, 12, 56 (notes); historical geography 11–23; National Delimitation 5, 29; political geography 23–31; physical geography 23–31
China 10–24, 32–36, 47, 54, 75, 84, 106, 120, 137, 150–57, 172, 182, 201; People's Republic of (PRC) 29, 31, 39, 44–54; western 81, 140–49
Chinese expansion 24–26; in Tarim basin 14, 49
Christianity 33, 38
climate 8–10, 80, 114, 116; micro- 52
collectivisation 46, 54, 94
commodities 47, 49, 110, 179
communications 48, 50, 67–69, 221
Communist 29, 46, 50, 54, 77, 106
consumption 71, 74, 78, 93–99, 110–11, 126–37, 111
cotton 52, 71, 85, 168–70, 178–81, 186, 189, 211
counter-development 72–73
crops 95, 138, 168, 180–81, 197; cash 68
Cyrillic script 43–44

Dalai Lama 24, 29, 36
demographic 46, 51, 186, 188, 193
desert 10, 13, 51, 105, 128, 142, 149, 161
desertification 32, 52, 77, 140–49, 198
developing countries (world) 72, 75, 78, 85, 93
development 54–56, 65–74, 94, 139, 170, 195–207; plan 170–77; policies 54–55, 215; programmes 51, 55, 86–87, 100, 123, 212; sustainable 56, 73, 86–87, 94, 98–100, 105, 115, 118–24, 140, 299, 208
dogs 134–35, 152

dunes 128, 143–48
dust 52, 160–64
Dzungaria 14, 25

economic theory 75–85, 98, 105, 167, 229
economy, command 105, 114; local 65, 72; market 105–116, 137–38; modern 44; nomadic 44, 105, 108–116
ecosystems 51, 97, 102, 118–21, 128, 137–38, 145, 200
environmental benefits 143; conservation 98; costs 97, 135; damage 52–55, 176; degradation 54, 66, 75, 79, 95–98, 130–36, 162, 209; economics 136; economists 79; impact 138, 156; problems 51, 80, 140, 159, 161, 166; protection 76, 86, 194, 214, 224
energy, solar 72, 85; sustainable 128
erosion 72, 137–38, 146, 197; soil 52–53, 95, 102, 128–30, 198–99; water 140–43; wind 140–48
Eurasia (Eurasian landmass) 3–11, 44, 55
exports 71, 82–84, 111, 178–79

farming 10, 44–45, 66–76, 81, 145, 150–57, 178, 180–81, 193, 297
fencing 132–33, 145, 149
Ferghana 17, 18
fertiliser 52, 71, 95, 127, 132–36, 157, 179, 181, 189
finance 68
food 46, 67, 93, 98, 112, 128, 147, 178–82, 187, 195–97, 211
forest 8–9, 44, 122, 129–37, 197–99, 205, 209

Ganges 16; Gangetic plain 18
Gansu 10, 23, 34, 48, 130, 135, 142, 152
gas 82, 113, 159, 169–184
Genghiz Khan 20–23, 34, 48, 98
globalisation 65–73
Gobi 9, 12, 55
Golden Horde 21
grain 53, 94–95, 125–36, 147, 178–79
grasslands 8, 53, 76, 126–36, 148, 168, 192
grazing 76, 129–34, 142–48, 151–55
Greek coinage 49; colonies 14; deities 35; language and script 40–41; rule 25, 45
Greeks 23
gully, gullying 129, 146, 149

Hazaras 23
Hephthalites (White Huns) 17, 36

Himalayas (Himalayan range) 10–11, 55, 126, 196–207
Hindu Kush 11–22, 48, 196–207
horse 9, 45, 47
Huns, White – see Hephthalites
hydroelectric 72, 127, 138–39, 188, 195, 199

Ili 14, 27, 42, 160–64
immigration 54, 186
imports 83–84, 136, 168–69, 176
India 12–24, 27, 30, 49–50, 67, 135, 201, 208–29
Indus 10–11
industrialisation 53, 65, 69, 80, 85, 105, 107, 168, 210
industry 46, 53, 71, 81–83, 96, 156, 180–82, 190, 213–16, 222, 230–32
information 72–73, 77, 99, 122, 135, 154, 200, 207, 212, 218, 232
infrastructure 67, 71, 85, 125, 170, 182, 187, 216, 221, 223, 233
Inner Asia 6
International Union for the Conservation of Nature (IUCN) 101, 122
investment 82–87, 95, 127, 135, 138, 169–70, 178–82
Iran 12, 14, 22, 30–33, 50, 170–72
Iranian nomads 13, 15, 31; peoples 14–23, 32, 37; plateau 11, 48
irrigation 10, 52, 85, 95, 128, 138–38, 162, 190, 197
Islam 19, 32, 36–39, 45

Jehol (Manchuria) 20
Jews 35
Judaism 20, 33–34

Kalmykia 23
Kara Kum desert 9, 11; canal 181
Karakhanids 19–20
Karakorum mountains 11, 16
Kashgar 19, 48
Kashmir 11, 30–31, 65
Kazakh Hordes 24–25; steppe 9, 25–26
Kazakhstan 4–5, 9, 33, 39, 46, 49, 50–54, 82, 84, 106, 159–66, 168, 182
Kazakhs 22, 27, 38, 47, 155
Khalkas 24–25
Khazars 20
Khojand 15, 20, 48
Khyber Pass 11
Kitans (Kara Khitais) 20
Kopet Dag 11–22

Kunlun Shan 10
Kushan 16–17, 35–36, 41, 49
Kyrgyz 42; Yenisei 19, 34
Kyrgyzstan 4, 18–19, 33–38, 50–52, 84, 106, 186–89
Kyzyl Kum desert 9

Ladakh 10, 48, 65–68
Lake Baikal 119–21
Lake Balkhash 17, 24, 48, 160–66
Lamaism (Lamaist Buddhism) 36–37, 46
landlocked 55, 86, 105, 115, 167–68
Latin script 43–44
Lhasa 25, 84, 125–27, 134, 138
livestock 47, 52, 95–101, 107–116, 126–35, 148, 150–53, 189, 197–98, 221, 224
local community 67, 72, 133; economy 68–7; industry 83, 85; initiatives 73; resources 68–70
Lop Nor 9, 35, 53

Manchu 24–28, 110; script 42
Manchuria 9, 14, 18, 23
Manichaeism 19, 33–34, 42
manufactured goods 47, 71
markets 197, 232; distant 68; domestic 167, 222, 230–31; export 74, 177, 222; global 71, 85–86, 167; local 83–85; world 83
Marxist 30, 39, 107, 116
Mediterranean 40, 47, 50, 70
Merv (Mary) 11, 27, 33–38
Middle East 7, 15, 20, 47, 50–51, 115
migration 10, 12, 22, 47, 106, 115, 187, 193, 197, 209, 220
military 12, 46 53, 93, 198
milk 72, 84, 96, 112–13, 178
mining 160, 182, 215, 221
modernisation 47, 52, 54, 70, 106
monasteries 46, 115
Mongol conquest 37; empire 20–22; peoples 8–23, 31–32; script 42–44
Mongolia 4, 7, 19–25, 30–31, 42, 84, 93–103, 104–17, 118–24; Inner 4, 25, 28, 31, 46, 53, 141–147; Outer 4, 25, 28
Mongolian People's Republic (MPR) 28, 39, 43, 46, 50–54
Mongolian plateau 9, 19, 22–23
Mongols 20, 82, 104–17
mountain development 192–94, 195–207; regions 186–94; states, kingdoms (Himalayan) 10, 27, 31, 52, 54
Muslim 8, 19, 27, 37, 51,

Nepal 10, 200
Nestorianism 23
Nisa 15
nomadic civilisation 82, 102; herders 10, 51, 81, 94; society 85, 104, 110
nomadism 14, 46–48, 106–14,
nomads 13, 45, 47, 54, 110–13, 130
nuclear 53, 71, 159–60

oases 8–10, 45, 142, 149
oil 68, 78, 82, 93–94, 159, 169–83
Oirats 24–26
Okhotsk, Sea of 8–9
Orkhon 17, 21, 42
overgrazing 53, 81, 96, 144–45, 189, 198
overpopulation 66

Pakistan 11, 30, 36, 172
Pamirs (Pamir range) 11–16, 48, 162, 190
Pamiri peoples 38
Parthia 15–16, 34, 49
pastoralism 45–47, 80, 114, 126–27
pasture 76–77, 96, 99, 131, 139, 152, 156, 168, 189–94, 198; degradation 53, 128–29
permafrost 8
pipelines 171–78, 182
pollution 53, 65–72, 77, 93–97, 134, 162, 191, 215, 223
population control 94, 193; density 8, 53, 132–33, 187, 196; growth 102, 145, 184, 186–89, 193–97
privatisation 109, 112, 116, 174, 175, 184
programmes, aid 71; development 86, 212; training 108, 125, 152–54, 205–6; stabilisation 175

Qing 24–27, 46
Qinghai 23, 81, 126, 135, 142

radioactive 96
raw materials 47, 96, 218, 222, 231
reform 84, 112–13, 126, 154, 170–84
reindeer 8, 44
religion 32–38, 55, 98
resources, mineral 93, 115, 167; natural 86, 126–28, 133–37, 190, 196–97; water 120, 143–45, 181
rivers 8–10, 45, 130, 142–43, 149
rodents 53, 129
Rome 15, 49
rural areas 86, 127, 135, 219; population 86, 212, 220; way of life 65, 76,

Russia 12, 24, 50, 119–20
Russian expansion 26–27, 30; Federation 23, 30–31

Sakas 13, 16, 41, 45
Samarkand 20–23, 33, 35, 48
Samoyedic 8
Sassanian (Sassanid) 17, 36, 53
scripts 39–44
sedentarisation 46, 131
sedentary 13, 19, 32, 45–51, 81, 105, 109–15
semi-desert 19, 44, 161
Semipalatinsk 53, 160
shamanism 31–32, 39
sheep 96, 111, 113, 150–56
Siberia 4, 9, 23, 25, 31, 44, 120
Sikkim 10, 126
Silk Roads 32, 34, 47, 51
social costs 95; problems 68, 196
socialist 52–54, 75, 110, 114–1
Sogdian script 41–42
Sogdiana 13
Sogdians 32, 45
soil 52–53, 81–85, 95, 99, 128–32, 148, 161, 189–90, 197; conservation 73
Soviet Central Asia 5, 11, 30, 39, 52–53, 192–94
Soviet Union, collapse of 29, 50, 115, 194
steppe 9–20, 25, 29, 32, 37, 42–47, 51–55, 105–6, 122, 142–46, 161
subsidies 71–72, 127, 135–36, 181, 218
Syr Darya 10, 52, 162
system, capitalist 77; centralised 71, 85, 116; communist 77; free market 105–9; socialist 108

taiga 8–9
Tajik 42
Tajikistan 4, 15–16, 30–32, 50, 52, 186–94
Tajiks 23
Taklamakan 9–10, 48, 141
Talas (Kyrgyzstan)18
Tamerlane (Timur) 21–22
Tarim basin 9–25, 33–37, 42–49; Darya 10
Tartary (Tartarie) 5–6
Tashkent 16, 20, 27, 48
technology 53, 65, 81–85, 96–102, 132–33, 150–57, 196, 205, 214–16, 221; capital-intensive 80–83; labour-intensive 81–83; industrial 51; sustainable 100

Tengrism 32, 39
Tibet 4, 25–31, 36, 39, 41–45, 53–54, 84, 125–39
Tibetan empire 23; plateau 10, 23, 126–30; script 42
Tibetans 13, 82
Tien Shan 10–19, 33, 47–48, 151, 162, 165
Tocharian 14, 41
tourism 137, 195, 197
toxic 52, 96–99
trade 46–50, 65–72, 107, 135, 137, 176, 196, 209
traditional culture 65; lifestyles 94, 102; practices 95–101, 110, 126; society 54–55, 82, 86
transhumance 10
Transoxiana 6, 15–22, 27, 33–37, 42–48
transport 45–46, 51, 71, 79, 127, 130–35, 166, 171, 193, 211
transportation 68–69, 85, 96, 172–78
Tsarist 27–28, 38, 46, 50, 106
tundra 8
Tungusic 9, 14, 20, 23
Turfan 34
Turkestan 5, 21, 25–27, 37; eastern (Chinese) 5, 10, 13, 25, 50; western (Russian) 5, 10, 27, 29, 50
Turkestan Autonomous Soviet Socialist Republic 5, 29
Turkic Khaghanate 18; peoples 8–25, 31–34, 37
Turkmen 27, 42, 171–84
Turkmenistan 4, 15–18, 38, 40–44, 50–52, 82, 167–84
Turks 12
Tuva 28, 49
Tuvinians 23, 25, 37

Uighurs 19, 22, 27, 34, 41; script 41
Ulaanbaatar 84–85, 104–113
Ural mountains 9, 13; river 9, 159
urban areas 79, 135, 187; crime 107; growth 134, 139; pollution 134
urbanisation 45, 105, 193
Ussuri 120–21
Uzbek 42
Uzbekistan 4, 13–18, 23, 32, 36–44, 50–52, 106, 186–89
Uzbeks 22, 47

valleys 129–30, 145, 165, 186–87
vegetation 8, 10, 126, 134, 143–47, 161–65

Volga 4, 8–9, 21–23

waste 70–72, 78, 93–101, 134, 159, 165
water 8, 10, 52–53, 76, 84, 93–102, 120,
  138, 145, 148, 161, 179–81, 188, 191,
  198–99, 221
women 66, 155, 199
wool 52, 96, 112, 150–56

Xinjiang 4, 19, 25–31, 36–39, 44, 47,
  50–54, 84, 142, 150–55

Xizang 4, 29, 47, 200

yak 10, 70
Yakuts 23, 38
Yuëh-Chih 14, 16
Yunnan 23, 130, 143

Zerafshan 13, 32
Zoroastrianism 31, 33, 37